Each week a different specialist shares their marketing secrets.

- **Get the inside track.**
  Each week Charlene digs into one of the 148 tactics profiled in this book or a new tactics that she's discovered. You get the inside track on tips and results from experts in that week's featured marketing method.

- **Keep on top of new and emerging tactics.**
  You'll hear from the early adopters of new marketing tactics so that you can learn from their successes and decide what's right for your business.

- **Anytime Access.**
  Be on the live call to get your questions answered, but if you can't - each call is archived for later listening.

3 Sessions FREE
Enter this code
148-0102

www.MarketingTacticTuesdays.com

Please share your tactics ...

There are hundreds of tactics and thousands of integrated combinations that you can use to get your message to market. I have done my best to include the most notable at the time of publishing, but the list is in no way complete.

I'd love to hear from you!

If you know of an advertising or promotional tactic not listed in **148 Ways**, *or if you would like to add to any of the profiles, please email* **148Ways@charlenebrisson.com**. *I may include your suggestion in the next edition with a mention of your contribution. It may also appear on the www.MarketingTacticTuesdays.com members website.*

All the best in your marketing success,

Charlene Brisson
The 3-Step Marketing Pro

148 Ways

to Advertise & Promote your Business

by Charlene Brisson, MAPC

Cover Photo - Las Vegas General Store, November 2009.

Order this book online at www.148WaysToAdvertise.com
or 3-StepMarketingPro.com

© Copyright 2010 Charlene Brisson
All rights reserved. No part of this publication may be reproduced, stored in a retrieval system, transmitted or distributed in any form by any means, electronic, mechanical, photocopying, recording, or otherwise, without the prior written or email approval of the author.

Warning and Disclaimer
Every effort has been made to make this book as complete and as accurate as possible, but no warranty or fitness is implied. The information provided is on an "as is" basis. The author and the publisher shall have neither liability nor responsibility to any person or entity with respect to any loss or damages arising from the information contained in this book.

Published by Charlene Brisson Communications.
ISBN: 1453868437

About the Author

Charlene Brisson, MAPC has over 25 years of global marketing and communication experience out of six countries. She started her career in media (radio, newspaper and magazines) and has worked with dozens of online and offline entrepreneur-driven corporations and helped hundreds of small business start-ups. Charlene's work has provided her the opportunity to create and manage hundreds of integrated traditional and new media marketing campaigns.

148 Ways to Advertise & Promote Your Business is a natural outcome of Charlene's real-life experiences. In fact, the majority of the 148 profiles were written from Charlene's direct hands-on campaign experience. She knows first-hand what the pros and cons are, because she's lived them. With the number of ways to reach customers growing almost daily, Charlene recognized the necessity of this guide to help marketers and business owners.

Known as the 3-Step Marketing Pro, Charlene has taken her experience and narrowed it down to 3 proven steps to help businesses succeed in generating sales by simplifying the process. These 3 steps as outlined in Chapter One are THE primary marketing principles of ALL successful businesses and have worked time and again throughout her career. The 148 tactics in this book are the tools needed to successfully execute Step 3.

As a lifelong learner, Charlene earned her Master of Arts Degree in Professional Communication (MAPC) from Royal Roads University in 2009 and an honours certificate in Internet Marketing from The University of British Columbia in 2002. She lives in Vancouver, British Columbia and works with businesses worldwide to optimize their marketing campaigns and increase sales.

Charlene Brisson is also the author of *16 Major Mistakes Marketers Make ... and How to Avoid Them*; host of the weekly webinar series *Marketing Tactic Tuesdays* as well as creator of the *Marketing Minute* video series that can be found on www.YouTube.com/3StepMarketingPro.

To book Charlene for your next event, visit www.3-StepMarketingPro.com.

In Praise of 148 Ways to Advertise

"Charlene has done it! She's taken 25 years of incredible marketing experience encompassing large corporations to the smallest entrepreneur startup and given us the techniques that she used to market them to the world. Whether you have lots of marketing dollars to spend or have a shoestring budget you will find useful tidbits to get your mind moving, your plans proceeding and your sales pouring in."

– Gary C. Bizzo
President, BC Urban Entrepreneur
Development Association

*"**148 Ways** is an excellent resource for both the newbie and experienced marketer! The list of tactics is huge with endless possibilities behind each example. The 3-Step Marketing Pro could not have made it any easier. Well worth adding to your business resource library."*

– Rebecca Happy
The Connector Gal and
Social Media Evangelist

"An expert who can identify 148 techniques for marketing and then summarize them in three basic steps is worth listening to. Charlene Brisson magically combines her mind, her imagination, and her experience into practical, useful steps toward marketing success. The ultimate result is the countless insights in these pages. Thanks to Charlene's genius, they will be priceless to you."

– Michael Real
Professor of Communication and Culture
Royal Roads University

*"**148 Ways** to Advertise & Promote your Business" and the 3-Step Marketing Model brilliantly demystifies and simplifies the daunting task of selecting effective and relevant promotional strategies from the sea of marketing options presented to today's business owner. As a business coach to start-ups I see this as a valued and much needed resource. Thanks Charlene!"*

– Lynne Brisdon, PCC
President, LivinginVision

In Praise of 148 Ways to Advertise

"Really successful business owners spend 80% of their time on marketing and delegate the rest. This book will masterfully keep you on track. Leave it on your desk and refer to it often. Whether you're brand spanking new or have been around the block, it's a must read. Charlene will transform your marketing perceptions and strategies."

– Shawne Duperon,
Six-Time EMMY® Winner
CEO, ShawneTV

*"**148 Ways** is the individual recipe book for the chefs of integrated marketing and a must have to regularly spice up your campaigns."*

– Phillip Stainton
Former Managing Director, Satchi & Satchi

*"**148 Ways** should be on the desk and computer of every entrepreneur and marketer, and a required reading for all business executives and marketing majors. Packed with experiential wisdom, it is a practical guide on how to take one's business to new heights."*

– Arvind Singhal, Ph.D.
Samuel Shirley and Edna Holt Marston Endowed Professor in the
Department of Communication, University of Texas @ El Paso

*"**148 Ways** is THE reference guide for a quick tip on ANY marketing idea you can employ. Perfect for all businesses large and small. A MUST GET book."*

– Pauline O'Malley
Author, Lifestyle Selling for Women
President, The Revenue Builder

"Ever felt stumped: What can I do to promote my business? Here's an amazing checklist of 148 tactics. Everything from Deal-a-Day to Wildpostings. If you haven't heard of these tactics (and 146 more!), pick up this book."

– Darcy Rezac, Judy Thomson and Gayle Hallgren-Rezac
Authors of WORK THE POND! Use the Power of Positive
Networking to Leap Forward in Work and Life

This book is dedicated to the memory of
Audrey Paterson, my dear friend and mentor.

Thank you to

Keith E. Jackson
Andre Phaneuf
Andrea Pasztor
Anne Pustil
Carmen Choy
Finnigan Brisson
Glenda Beaulieu
Iris Chen
Jonah Tabuena
Jonathan Cohen
Lorraine Blakeman
Madeline Brisson
Mark Brisson
Rebecca Happy
Richard T. Freeman
Ronalisa Co
Shapiro Cohen
Tania Music
TayTay Brisson

Thank you to all the dedicated marketers who
continue to stretch the boundaries
with their creativity and genius.

Table of Contents

About the Author 5

Introduction
    A Few Statistics 13
    A Rapidly Shifting Industry 14
    So Why a Guide for Marketers? 15

Chapter One
    The 3-Step Marketing Model 19
        1. Know your IDEAL customer 20
        2. Craft a message that speaks to
            your IDEAL customer 26
        3. Select effective marketing tactics to reach
            your IDEAL customer 29

Chapter Two
    The Complete List of 148 Tactics 35

Chapter Three
    Offline Tactics at-a-Glance 45

Chapter Four
    Online Tactics at-a-Glance 53

Chapter Five
    Mobile Tactics at-a-Glance 59

Chapter Six
    148 Tactic Profiles 63

Chapter Seven
    Marketing Support 379

*In 2008 more than $142 billion
was spent on advertising
in the United States.*

Photo at right:
Taken by Charlene Brisson
in New York City, 2006.

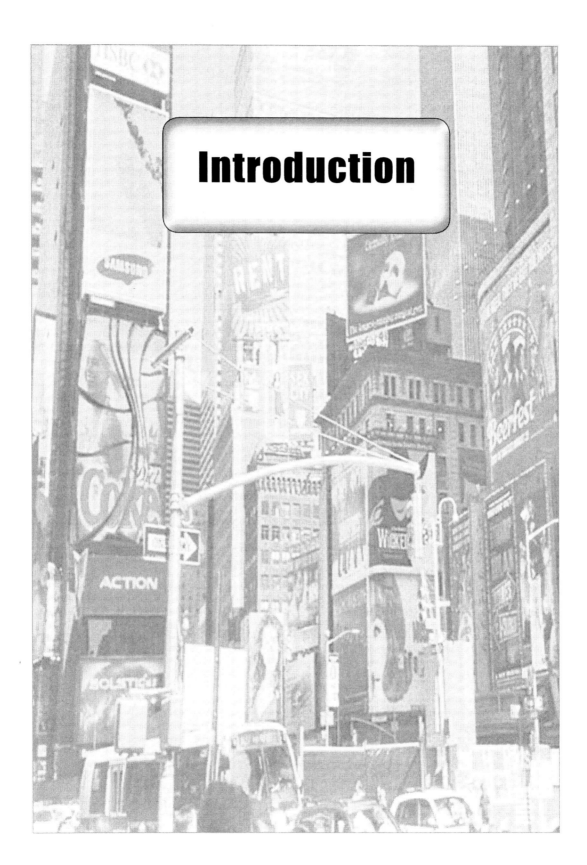

The business of marketing is changing so rapidly, that keeping up-to-date on what's new can be an enormous challenge.

Introduction

Without a doubt, marketing is one of the most exhilarating industries on the planet. It directs the pulse of the multi-billion dollar consumer industry and influences in more ways than are imaginable. The explosive growth of the internet and mobile phone accessibility have ramped up the pace and put additional stress on business owners and marketing professionals to keep ahead of the competition.

At this writing, not even the global recession has affected advertising revenues in a significant way. Overall, the industry has suffered only slight losses between 2007 and 2009.

Here's a few statistics …

Ad Age reports that in 2008 more than $142 billion was spent on advertising in the United States – $9.73 billion of which was spent on the internet. Although down in single digits from 2007 due to the general economy, the numbers continue to be extraordinary[1].

In the UK 18.6£ billion was spent on advertising in 2008 – 19.3% of which was on the internet (internet is up 9% since 2006)[2].

The Direct Marketing Association claims that in 2008, marketers — commercial and nonprofit — spent $176.9 billion on direct marketing in the United States, generating approximately $2.057 trillion in incremental sales[3].

---

1 http://adage.com/images/random/datacenter/2009/spendtrends09.pdf
2 http://www.adassoc.org.uk/aa/index.cfm/adstats/
3 http://www.the-dma.org/cgi/disppressrelease?article=1228

A Rapidly Shifting Industry

Although it is true that some mediums like newspapers and television networks have been hit with debilitating losses, these losses are being made up by record increases in online marketing expenditures. The media consumption habits of consumers are forcing corporations to shift from heavy traditional marketing to shared budgets with online and wireless media.

Over the past ten years marketers have watched the industry change dramatically. Newspapers in particular, have been dropping readers for decades and are only recently paying the full price for not finding new ways to meet the needs of the modern information consumer. In fact, several of the oldest newspapers in North America have closed their doors. Along with dozens of others in 2009 that traded their print editions for online issues were The Seattle Post Intelligencer and the 174 year old Ann Arbor News. Colorado's Rocky Mountain News shut their doors completely, after being published since 1859. Although shocking to many, it doesn't mean that the advertising industry is slowing down or that valuable marketing tactics are being lost. In fact, these tactics are simply being replaced with new methods of reaching the target. Consumers are demanding their information in new ways and companies unable to adjust are being left behind.

There's also a lot of talk about how online marketing is replacing traditional marketing activities like newspapers, radio, outdoor, and the like. This is similar to how TV was foretold to obliterate radio and online was going to render magazines useless. Quite the opposite is happening. The internet has merely spurred new opportunities and provided marketers with a more holistically integrated approach. When you take a look at the list of tactics in *148 Ways*, you too will see that as long as there's a consumer and a space where they frequent, there will always be offline ways of reaching them. A huge vote of approval for integrated online and offline campaigns is that large online marketers like Amazon.com, Zappos.com and Ally.com are using offline tactics to expand their brand awareness[4].

So Why a Guide for Marketers?

This guide bears witness to the sheer genius and willpower of marketers to find new and creative ways of getting their message to market. From a practical perspective, it also provides a necessary tool for anyone that has a product or service that someone out there wants, or better yet, needs. If you're reading **148 Ways** and you know who your target market is, there is no reason why you can't succeed in reaching your audience with a combination of the tactics found within these pages – *with or without a budget*. And if you don't know who your target market is, read Chapter One very closely.

Yes, that's right, the first step to EVERY successful campaign is to know who your customer is – also referred to as market, target market, audience, prospect, customer profile, client, segment and/or niche. It is only once you've identified the demographics, geographics and psychographics of your primary customer that you can then craft your message and finally identify an integrated group of effective tactics to deliver that message. So in addition to a robust list of tactics, this guide is also a reminder to marketers that there is an important step before selecting tactics. Once you have identified your IDEAL customer, the tools and tactics at your disposal are listed in the following pages. They are the methods that marketers use every day to connect goods and services to the millions and millions of people who need them.

---

4 http://www.adweek.com/aw/content_display/news/agency/e3i-d499f8aa1018de8318bc17705265119c

More formally, *148 Ways* has been written specifically for these three reasons:

1. **To provide a simple guide** for business owners and marketers to look beyond existing advertising and promotional campaigns - to help you step outside of your box and integrate marketing methods, techniques and tools that you may not know about or have thought of. We just don't know what we don't know, even if we've been in the business for a long time. And the business of marketing is changing so rapidly, that keeping up-to-date on what's new can be an enormous challenge, if not impossible.

2. **To help business start-ups identify what will work.** Having counseled hundreds of new business start-ups, I observed that it is the marketing component that is often the most daunting part of the business plan. It is also the most important as marketing comprises at least two-thirds of the plan. If there's no plan to get the product into the hands of buyers – there's no sales and we all know what that means – no money.

3. **Information is power.** *148 Ways* is the ultimate educational tool for those learning and entering the industry for the first time. Having taught advertising courses, I've seen first hand the glaring lack of compiled tactics. It has taken two dedicated years to create this list, and I'm in the business. Knowing what advertising tools and tactics are out there and which ones to use for what purpose is difficult. This guide is practical and an easy reference. It is the perfect complement to theory and everything that a textbook is not.

148 Ways is all about tactics. Tactics are vital to developing an effective marketing campaign. There are two additional elements that come before selecting tactics that are also critical. Chapter One provides a "starter" on the 3-Step Marketing Model to show you how to profile your IDEAL customer (Step 1); craft a message that speaks to your IDEAL customer (Step 2); and give you a system to select effective tactics that will reach your customer (Step 3).

Chapter One
- 3-Step Marketing -

An effective campaign comprises three critical components: the right message to the right audience using the right medium.

The 3-Step Marketing Model

This book is a guide of tactics to use in getting your message to market. In itself, it provides numerous ways to reach your customer. However, for any of the tactics to work, it has to be a reality within your customer's day-to-day activities. An effective campaign comprises three critical components: the right message to the right audience using the right medium. Without these three you may as well go to the casino and put a big fat wad of cash on Red 17 at the roulette table and hope for the best.

For example, there's no point advertising on radio if your target market doesn't listen to the radio, or online if your customers aren't computer literate and don't own computers. It's all about finding the direct route to your customer through these three proven simple steps to successful marketing.

1. ***Know*** your IDEAL customer.
   (also known as market, target market, client, prospect, customer profile, audience, segment and/or niche.)

2. Craft a message that ***speaks*** to your IDEAL customer.

3. Select effective marketing tactics that ***reach*** your IDEAL customer.

These are the steps that I have used for decades to build successful campaigns. I've discovered that no matter how technology changes, these 3 primary marketing principles never change and they will always work for you.

1. Know your IDEAL customer.

The hub of every successful campaign, and in fact, the foundation of all successful business is having a thorough understanding of your customer. Without knowing who is the most likely individual to buy your product or service there is no way for you to produce a responsive or cost effective campaign. You will be destined to waste a great deal of time and money. It is only once you've identified your customer that you can then craft a resonating message, and ultimately select the marketing tactics that will reach them. You can't be everything to everyone, so the more narrow and focused you target, the more successful your campaign will be.

> Let's say you sell **red stoves** and have a client profile defined as female, 35-55, living in Los Angeles. It may sound good, but the definition is so broad that you will have difficulty zeroing in on how to find her. In fact, the 2000 US census indicates that women in this age group comprise 31.6% of LA. Now that's daunting. So, the more information you can gather to narrow in on where and who she is, the easier and more likely you will be able to find her and get her attention.
>
> For example, let's say that this **red stove customer** is an urban dweller rather than a suburbanite. Through some basic research, you also discover that she has a household income of $65,000+, listens to the radio and grocery shops at the local organic chain store like Whole Foods Market. In addition, she is moderately computer literate and is predisposed to researching all of her major purchases online before buying. She keeps up with extended family through Facebook.

Every bit of this information extracted about your "IDEAL" customer provides significant information to create a responsive campaign.

Marketers speak of target markets in terms of geographics, demographics and psychographics. Each area provides clues on how to find and communicate with your buying client.

Here's a business-to-consumer (B2C) guide that will help you do a thorough analysis of your customer. See how many answers you can provide. Keep in mind that these questions may not all apply, but the more specific you get, the more targeted and effective your marketing will become.

- **Geographics** – where your customer is located.
    - Country
    - Region
    - Province/State
    - City/s
    - District/Community/s

- **Demographics** – basic facts about your customer.
    - Age
    - Gender
    - House Hold Income
    - Education
    - Job/Industry
    - Ethnicity
    - Marital Status
    - Number of Children
    - Age of Children
    - Home Ownership/Rental
    - Religion

- **Psychographics** – lifestyle behaviors as they pertain to buying decisions.

    - Is a member of these associations
    - Belongs to these clubs
    - Reads these magazines
    - Reads these newspapers (community, daily, local, national, free, paid)
    - Listens to these radio stations
    - Watches these television shows
    - Dines at these restaurants
    - Shops for food at
    - Shops for clothes at
    - Drives this make of vehicle
    - Owns these pets
    - Has this kind of personality
    - Likely to vote for this political party
    - Volunteers for this type of organization
    - Donates to these causes
    - What else is unique to your IDEAL customer?

- **Online Psychographics** – online and wireless lifestyle behaviors as they pertain to buying decisions.

    - Has this level of computer skill
    - Regularly visits these information sites
    - Listens to these kind of podcasts
    - Watches these kind of videos/vodcasts
    - Uses this type of smartphone/iPhone/cell phone
    - Has downloaded these kind of applications
    - Belongs to these social media groups
    - Engages in these mobile activities (texting, browsing, payments, scanning)
    - Purchases these kind of items online

If you can answer the previous questions, you will be provided with a treasure trove of clues to find your customer and know how to speak directly to them.

Information Overload

At this point, you're probably thinking ... where am I going to find all of this information about my customers? Good question! There are four ways that I approach building the information base for customer profiles. Keep in mind that you're looking for patterns of similarity so that you can group your audience into segments that will respond to like communications and marketing efforts.

1. **Ask your customers**. Engaging with your clients face-to-face through customer service interaction; while providing your service; selling your product; or while they are browsing; are terrific opportunities to ask questions. Remember, in general, people love to talk about themselves. Keep a few questions on hand that you and/or your team can ask without being intrusive and try to work them into conversations. For example, if you sell gardening supplies it's perfectly in-line to ask prospects and customers how often they garden, what do they garden and how big is their garden. None of these questions would be considered intrusive based on the product AND the answers will provide you with useful profiling information. Asking for an email address when confirming the order will provide you with a future follow-up opportunity plus give you an indication of how many of your customers are online. People will willingly provide this contact info if you position it that you would like to send them updates

on new products and upcoming sales (there has to be something in it for them). Also make assurances that their email address will be kept confidential and won't be shared with any other company.

2. **Observe your customers.** Watching for behaviors and listening to what your customers say can provide a tight perspective of who they are. Observe what they're wearing, their grooming, clothing, accessory style, jewelery (wedding ring), hair, makeup, fitness level and maybe even what kind of car they drive. It's also easy to observe demographics such as age, general income and ethnicity. You can translate these observations into behaviors. For example, if your client's hair is always looking fabulous - great color, no roots and modern style, you can jump to the assumption that she goes to a hair salon to have it done. Going to a hair salon is a lifestyle behavior that is important in profiling your customer. Based on whether or not she works, what her basic household income is, her general age and where she lives, you can then narrow down which salons she (as a group) is likely to frequent. For online customers, observe their site and click through behaviors. You'll need this kind of information in Step 3 to connect the dots to effective marketing tactics where it's discussed in more detail.

3. **Do research.** Census statistics can provide an abundance of information about your customers. If you have a couple of basic demographics like gender and household income (HHI), you can cross reference and narrow down what primary neighborhoods your customers live in. If you have gender, number of children and neighborhoods, you can cross reference their household income. More

advanced research can be conducted by companies whose service it is to map your customers and provide detailed segmentation information and commonly used segment names. These companies can also provide customer target locations right down to postal walks. Ad agencies often use this profiling research for large clients. This research is quite costly and requires a robust database of existing customers.

4. **Conduct surveys.** Another way to capture information on your customers is to run surveys. For an offline business without a customer base, craft a short survey of 10 questions, stand outside the doors of your competition (off to the side, of course) and ask people who walk out to engage in your survey. You can't get more targeted. Online surveys are very easy to produce and many are free like http://www.surveymonkey.com/ and http://www.kwiksurveys.com/. These surveys are templated to make it really easy and can be sent by email to a bulk list or designed as pop-ups on your website. When emailing, the data you get from your surveys will only be as good as your list, so make sure you are sending the survey to a recent list of very targeted individuals. For businesses that already have history, your customer database should be a good list if you've been keeping it up-to-date. Also keep in mind that many people block pop-ups, so this type of survey on your website may retrieve questionable data.

2. Craft a message that speaks to your IDEAL customer.

Once you get clear on your target market, you'll have a greater understanding of how to address them in a way that will grab their attention.

And what is it that your target market wants to know? More than anything, they want to know these two things …

Why should I do business with you rather than anyone else?

AND …

What's in it for me?

These two questions translate into the necessity of sharing 1) your unique selling position (also known as your value proposition) and 2) the benefits of your product/service which will show how you're going to solve their problems.

So why should someone do business with you?

A basic marketing principle is to clearly identify what differentiates your product/service from the competition. By doing so, you communicate to your target market why they should do business with you rather than anyone else.

This is called your unique selling position (USP), and it is what will really set you apart. It is what you do that is bigger, better, faster or stronger than anyone else. "Better quality" doesn't count – everyone says better quality – **prove it**. Your USP must be tangible and show value.

> Let's use **Red Stoves Inc.** as an example again. If you Google "**red stove**" (in quotes), 36,200 listings appear. This tells us that selling red stoves is not all that unique. But what the business does which is completely different than all other vendors and of great value to the customer is that **Red Stoves Inc.** offers free delivery 24/7. Now that's a unique selling position - a value proposition that will get the customer's attention.
>
> This USP addresses the customer's pain of having to take time off work (and often losing pay) to be there when the delivery is made.

And what's in it for me?

Over and above the USP, your product/service has several benefits to the customer that you may also want to communicate. It helps to create a chart - to make a list of FABS (features and their associated benefits) for your product or service. **A feature is a fact. A benefit is a personal gain.**

When creating your list, keep in mind that people are wanting you to solve their problems and resolve their pain. How do your benefits do this?

Here's an example of features and benefits (FABS) using **Red Stoves Inc.**

Feature	Benefits
Red Colour	1) Excellent resale value. 2) Gives owner prestige as friends look on with envy.
Energy Saving	1) Money saving. 2) Being and being seen as environmentally concientious.

Do you see how the benefits address the customer's pain of money issues and the importance of how they are seen by others?

The benefits of your product/service are what will comprise your messaging and advertising copy. It's the "What's in it for me" for your target customer.

3. Select effective marketing tactics that reach your IDEAL customer.

Now that you've identified who your target market is and clarified your USP and FABS, it's time to figure out what tactics you're going to use to get your message out.

This is where you can waste a great deal of money using mediums and tactics that don't work to attract your specific customer.

Saving time and money is why it's so important to do a complete customer profile of geographics, demographics and psychographics - so that you can strategically and effectively target to get the optimum response.

In addition to not knowing who your customer is, there can be two major issues here. *First*, many business owners make the mistake of using advertising vehicles that THEY enjoy themselves. Remember, *it's not about you*, the business owner. It's about your customer and only your customer. If the tactic you're using isn't aligned with the day-to-day life of your customer, then it's not right and your efforts will be ineffective. *Second*, is simply getting overwhelmed. Many business owners don't know what mediums and tactics are out there to use. And with all the talk about online marketing and social media like Twitter and Facebook, it can get really complicated, really fast. This is another reason why your customer profile is so important - these clues will lead you directly to the right marketing tactics.

The jump from knowing who your customer is, to what tactics to use, is like connecting the dots. As discussed earlier, there's no point advertising on radio if your target market doesn't listen to the radio, or online if your customers aren't computer literate and don't own computers.

Here's an example using **Red Stoves Inc.** and going back to the customer profile we discussed earlier.

> We can loosely define the **Red Stoves Inc.** client profile as female, 35-55, living in urban Los Angeles. She has a household income of $65,000+; listens to the radio; grocery shops at the local organic chain store; is computer literate and likes to research her purchases. She keeps in touch with extended family through Facebook.
>
> If you look closely, this information provides you with clues to know what tactics to use to reach her and how to speak to her. Here are some of the dots connected …

Characteristics / Clues	Tactics
Listens to radio + Female 35-55	Local Radio Campaign – on station that women 35-55 are primary listeners.
Organic shopper + Urban dweller	Partnership with popular organic grocery store/ chain in urban communities.
Computer literate + Researches online + Facebook	Website Pay-Per-Click Organic search engine optimization (SEO) Social Media Online contest

Integrate tactics for optimum results.

Your best results will always come from combining the right tactics with consistent messaging. Integrated campaigns along with consistency of message and look are critically important. Every time **Ms. Red Stove Buyer** sees or hears from **Red Stove Inc.** she must experience a repetitive impact – no matter what the medium – radio, online, email, print, outdoor, etc. By the 7th to 20th time of seeing similar creative, she will eventually start paying attention and then make a buying decision if she's in the market. Here's a general idea of how to use the target market profile and tactics to build an adhoc (also known as one-off) campaign to reach **Ms. Red Stove Buyer**.

> **Red Stoves Inc.** can do a partner contest with Organic Grocery Ltd. The idea is to give away one **red stove** each week for four weeks to Organic Grocery customers. This strategically makes sense because Organic Grocery Ltd. has primarily the same customer base that **Red Stoves Inc.** has. Remember the customer profile – partner with a ready-made audience.
>
> Here's how it goes. Everyone who makes a purchase at Organic Grocery gets an entry form handed to them personally at check out. The live script is – "We're giving away four **red stoves** this month. When you enter online, you'll also get a $5 coupon for the next time you shop at Organic Grocery." The coupon gives customers a second reason to make the effort to enter online - 1. to win the stove and 2. a discount when they return to shop.

During the contest, Organic Grocery Ltd. advertises the **Red Stoves** giveaways in their regular bi-weekly newspaper flyers; in their online newsletter; and within their regular weekly radio campaign.

Red Stoves Inc. adds to the radio budget to increase the number of giveaway ads that run, plus runs general awareness ads specifically promoting **Red Stoves Inc.**

Simultaneously, **Red Stoves Inc.** runs a robust geographically targeted online support campaign using Facebook ads, and pay-per-click that drive users to the contest entry site. The entry site is developed by **Red Stoves Inc.** (which gives them full control) and includes prominent exposure for Organic Grocery Ltd.

Red Stoves Inc. also creates a Facebook fan page where they post interesting content about their red stoves along with easy links for visitors to enter the contest.

To engage existing customers, **Red Stoves Inc.** launches an email campaign encouraging customers to enter online, join their facebook fan page and take advantage of the $5 coupon for Organic Grocery Ltd.

Keep in mind that entrants must answer a skill testing question to enter. In most locations, it's against the law to force someone to purchase in order to be entered into a contest.

This campaign works for both parties because Organic Grocery Ltd. is providing "added value" to their customers with an additional incentive to return. On the other hand, **Red Stoves** Inc. is getting exposure to a targeted audience. Both companies have access to the online buyers list generated from the online entries to be used for future marketing campaigns.

---

148 Ways to Advertise & Promote your Business presents a multitude of traditional, online and mobile media tactics on the following pages. Use this guide to connect your own dots from customer to tactics; as a tool to help you expand campaigns; to generate ideas; and to increase your general marketing knowledge.

You can waste a great deal of money using mediums and tactics that don't work to attract your specific customer.

*Photo at right:
Taken by Carmen Choy
in Hong Kong, 2010.*

Chapter Two
-148 Tactics-

If the tactic you're using isn't aligned with the day-to-day life of your customer, then it's not right and your marketing efforts will be ineffective.

The Complete List of 148 Tactics

Tactic .... **Page**

1. Aerial Banner .... Offline 65
2. Aerial Skywriting.................. Offline 67
3. Affiliates .... Offline .... Online 69
4. Airplane - Interior .... Offline 71
5. Airplane - Tray Tables.......... Offline 73
6. Airport Security Buckets ..... Offline 75
7. Arena Boards....................... Offline 77
8. ATM Machines .... Offline 79
9. Bags - Eco-Friendly .... Offline 81
10. Balloons .... Offline 83
11. Banner Ads .... Online......................... 85
12. Barcode Marketing.............. Mobile 87
13. Bathroom Ads .... Offline 89
14. Billboards............................. Offline 91
15. Billboards - Human .... Offline 93
16. Blogging .... Online......................... 95
17. Brochures............................. Offline 97
18. Building Wrap .... Offline 99
19. Bus Benches .... Offline 101
20. Buses..................................... Offline 103
21. Business Cards .... Offline 105

Tactic		Page
22. Bus Shelters	Offline	107
23. Catalogue	Offline .... Online	109
24. Coupon Books	Offline	111
25. Coupon Book - Entertainment	Offline	113
26. Coupon Envelopes	Offline	115
27. Cups & Napkins	Offline	117
28. Deal-a-Day	Online .... Mobile	119
29. Decals	Offline	121
30. Decals - Floor	Offline	123
31. Direct Mail	Offline	125
32. Directories	Online	131
33. Door Hangers	Offline	133
34. eBoards	Offline	135
35. Email - Auto-Responders	Online	137
36. Email - Blasts	Online	139
37. Embedded Logos	Offline	141
38. Embedded Logos - Food	Offline	143
39. ePostcard Xpress	Offline .... Online	145
40. Escalator Handrails	Offline	147
41. Escalator Steps	Offline	149
42. Ezine - Ads	Online	151

Tactic			Page
43. Ezine - Article Marketing	Online		153
44. Ezine - Publisher	Online		155
45. Facebook - Ads	Online	Mobile	157
46. Facebook - Fan Page	Online	Mobile	159
47. Facebook - Group	Online		161
48. Flyers	Offline		163
49. Gobos	Offline		165
50. Graffiti - Green	Offline		167
51. Grafitti - Live	Offline		169
52. Graffiti - Permanent	Offline		171
53. Grass Painting	Offline		173
54. Ice Cubes	Offline		175
55. Inflatables	Offline		177
56. Intranet	Online		179
57. iPhone/Smart Phone Applications (Apps)	Online	Mobile	181
58. LinkedIn - Ads	Online	Mobile	183
59. LinkedIn - Profile	Online	Mobile	185
60. Magazines	Offline	Online	187
61. Meet-ups	Offline	Online	191
62. Mobile Billboards	Offline		193
63. Mobile Marketing	Mobile		195

Tactic			Page
64. Mobile Video Cubes	Offline		197
65. Movie Ads	Offline		199
66. Murals	Offline		201
67. Name Tags - LED	Offline		203
68. Newsletter	Offline	Online	205
69. Newspaper - Commuter	Offline	Online	207
70. Newspaper - Community	Offline	Online	209
71. Newspaper - Daily	Offline	Online	211
72. Paid to Read (PTR)	Online	Mobile	215
73. Parking Stripes	Offline		217
74. Pay per Click (PPC)	Online	Mobile	219
75. Pedestrian Motion Panels	Offline		223
76. Placemats	Offline		225
77. Podcasts	Online	Mobile	227
78. Pop-up Ads	Online		229
79. Pop-up Store	Offline		231
80. Postcards	Offline		233
81. Posters	Offline		235
82. Pre-roll Video	Online		237
83. Prizing - Media	Offline	Online	239
84. Prizing - Not-for-Profits	Offline	Online	241
85. Product Placement	Offline		243

Tactic			Page
86. Promotional Items	Offline		245
87. Proximity Advertising	Offline	Mobile	247
88. Publicity	Offline	Online	251
89. Rack Cards	Offline		253
90. Radio - Commercial	Offline	Online	255
91. Radio - Online	Online		259
92. Radio - Online - Produce Your Own	Online		261
93. Radio - Satellite	Offline	Online	265
94. Referral Program	Offline	Online	267
95. Reviews and Testimonials	Offline	Online	269
96. Rewards Program - Build Your Own	Offline	Online	271
97. Rewards Program - Provide Product	Offline	Online	273
98. RFID - Billboards	Offline	Online	275
99. Sampling	Offline		277
100. Shopping Carts	Offline		279
101. Shopping Channel	Offline		281
102. Sign Spinning	Offline		283
103. Signage - Dynamic LCD	Offline		285
104. Signage - LED Text Only	Offline		287

Tactic			Page
105. Signage - POP LCD	Offline		289
106. Signage - Sandwich Boards	Offline		291
107. Signage - Smart Poster	Offline		293
108. Signage - Traditional	Offline		295
109. Solo Ads	Online		297
110. Speaking Engagements	Offline	Online	299
111. Sponsor a Team	Offline		301
112. Sponsorships	Offline	Online	303
113. Squidoo	Online		305
114. Street Teams	Offline		307
115. Subway Tunnel	Offline		309
116. Tattoos	Offline		311
117. Team Events	Offline		313
118. Teleseminars	Offline	Online	315
119. Television	Offline		317
120. Tent - Pop-up	Offline		321
121. Text Messaging	Mobile		323
122. Ticket Jackets	Offline		325
123. Trade Shows	Offline		327
124. Transportation Straps	Offline		329
125. Twitter	Online	Mobile	331
126. Vehicles	Offline		333

Tactic			Page
127. Video Games-Advergaming	Online	Mobile	335
128. Video Games - In Game	Online	Mobile	337
129. Video Media Releases	Online		339
130. Video Projection	Offline		341
131. Viral Marketing	Online		343
132. Vodcasts	Online	Mobile	345
133. Voice Blasts	Offline	Mobile	347
134. Water Bottles	Offline		349
135. Waterfall Logo	Offline		351
136. Webinars	Online		353
137. Website	Online	Mobile	355
138. Website Links	Online		357
139. White Papers / Reports	Online		359
140. Wikki Media Commons	Online		361
141. Wildpostings	Offline		363
142. Wildpostings - Rip Away	Offline		365
143. Window Art	Offline		367
144. Window Display	Offline		369
145. Word-of-Mouth	Offline		371
146. Yellow Pages	Offline	Online	373
147. YouTube - Ads	Online	Mobile	375
148. YouTube - Videos	Online	Mobile	377

137 Ways to Advertise and Promote Your Business

As long as there's a consumer and a space where they frequent, there will always be offline ways of reaching them.

*Photo at right:
Taken by Charlene Brisson
in Vancouver, BC, Canada
during the 2010 Winter Olympics.*

Chapter Three
-Offline-

By clearly identifying your USP, you communicate to your target market why they should do business with you rather than anyone else.

Offline Tactics at-a-Glance

This list of offline tactics is extracted from the complete list of 148 traditional and new media tactics on pages 37-43.

Tactic ....**Page**

1. Aerial Banner 65
2. Aerial Skywriting............................67
3. Affiliates 69
4. Airplane - Interior 71
5. Airplane - Tray Tables 73
6. Airport Security Buckets 75
7. Arena Boards 77
8. ATM Machines...............................79
9. Bags - Eco-Friendly........................81
10. Balloons 83
11. Bathroom Ads................................89
12. Billboards.......................................91
13. Billboards - Human 93
14. Brochures.......................................97
15. Building Wrap 99
16. Bus Benches................................101
17. Buses..103
18. Business Cards 105
19. Bus Shelters.................................107
20. Catalogue.....................................109

Tactic	Page
21. Coupon Books	111
22. Coupon Book - Entertainment	113
23. Coupon Envelopes	115
24. Cups & Napkins	117
25. Decals	121
26. Decals - Floor	123
27. Direct Mail	125
28. Door Hangers	133
29. eBoards	135
30. Embedded Logos	141
31. Embedded Logos - Food	143
32. ePostcard Xpress	145
33. Escalator Handrails	147
34. Escalator Steps	149
35. Flyers	163
36. Gobos	165
37. Graffiti - Green	167
38. Grafitti - Live	169
39. Graffiti - Permanent	171
40. Grass Painting	173
41. Ice Cubes	175
42. Inflatables	177
43. Magazines	187

Tactic	Page
44. Meet-ups	191
45. Mobile Billboards	193
46. Mobile Video Cubes	197
47. Movie Ads	199
48. Murals	201
49. Name Tags - LED	203
50. Newsletter	205
51. Newspaper - Commuter	207
52. Newspaper - Community	209
53. Newspaper - Daily	211
54. Parking Stripes	217
55. Pedestrian Motion Panels	223
56. Placemats	225
57. Pop-up Store	231
58. Postcards	233
59. Posters	235
60. Prizing - Media	239
61. Prizing - Not-for-Profits	241
62. Product Placement	243
63. Promotional Items	245
64. Proximity Advertising	247
65. Publicity	251
66. Rack Cards	253

Tactic	Page
67. Radio - Commercial	255
68. Radio - Satellite	265
69. Referral Program	267
70. Reviews and Testimonials	269
71. Rewards Program - Build Your Own	271
72. Rewards Program - Provide Product	273
73. RFID - Billboards	275
74. Sampling	277
75. Shopping Carts	279
76. Shopping Channel	281
77. Sign Spinning	283
78. Signage - Dynamic LCD	285
79. Signage - LED Text Only	287
80. Signage - POP LCD	289
81. Signage - Sandwich Boards	291
82. Signage - Smart Poster	293
83. Signage - Traditional	295
84. Speaking Engagements	299
85. Sponsor a Team	301
86. Sponsorships	303
87. Street Teams	307
88. Subway Tunnel	309
89. Tattoos	311

Chapter Three - Offline Tactics at-a-Glance

Tactic	Page
90. Team Events	313
91. Teleseminars	315
92. Television	317
93. Tent - Pop-up	321
94. Ticket Jackets	325
95. Trade Shows	327
96. Transportation Straps	329
97. Vehicles	333
98. Video Projection	341
99. Voice Blasts	347
100. Water Bottles	349
101. Waterfall Logo	351
102. Wildpostings	363
103. Wildpostings - Rip Away	365
104. Window Art	367
105. Window Display	369
106. Word-of-Mouth	371
107. Yellow Pages	373

The internet has spurred new opportunities and provided marketers with a more holistically integrated approach.

Chapter Four
-Online-

Each and every minute, 24-hours of video is uploaded onto YouTube.

Online Tactics at-a-Glance

This list of online tactics is extracted from the complete list of 148 traditional and new media tactics on pages 37-43.

Tactic	Page
1. Affiliates	69
2. Banner Ads	85
3. Blogging	95
4. Catalogue	109
5. Deal-a-Day	119
6. Directories	131
7. Email - Auto-Responders	137
8. Email - Blasts	139
9. ePostcard Xpress	145
10. Ezine - Ads	151
11. Ezine - Article Marketing	153
12. Ezine - Publisher	155
13. Facebook - Ads	157
14. Facebook - Fan Page	159
15. Facebook - Group	161
16. Intranet	179
17. iPhone/Smart Phone Apps	181
18. LinkedIn - Ads	183
19. LinkedIn - Profile	185
20. Magazines	187

Tactic	Page
21. Meet-ups	191
22. Newsletter	205
23. Newspaper - Commuter	207
24. Newspaper - Community	209
25. Newspaper - Daily	211
26. Paid to Read (PTR)	215
27. Pay per Click (PPC)	219
28. Podcasts	227
29. Pop-up Ads	229
30. Pre-roll Video	237
31. Prizing - Media	239
32. Prizing - Not-for-Profits	241
33. Publicity	251
34. Radio - Commercial	255
35. Radio - Online	259
36. Radio - Online - Produce Your Own	261
37. Radio - Satellite	265
38. Referral Program	267
39. Reviews and Testimonials	269
40. Rewards Program - Build Your Own	271
41. Rewards Program - Provide Product	273
42. RFID - Billboards	275
43. Solo Ads	297

Tactic	Page
44. Speaking Engagements	299
45. Sponsorships	303
46. Squidoo	305
47. Teleseminars	315
48. Twitter	331
49. Video Games - Advergaming	335
50. Video Games - In Game	337
51. Video Media Releases	339
52. Viral Marketing	343
53. Vodcasts	345
54. Webinars	353
55. Website	355
56. Website Links	357
57. White Papers / Reports	359
58. Wikki Media Commons	361
59. Yellow Pages	373
60. YouTube - Ads	375
61. YouTube – Videos	377

TV was foretold to obliterate radio and online was going to render magazines useless.

Chapter Five
-Mobile-

*As printed newspapers fall,
other communication mediums
like mobile phones are
taking their place.*

Mobile Tactics at-a-Glance

This list of mobile tactics is extracted from the complete list of 148 traditional and new media tactics on pages 37-43.

Tactic	Page
1. Barcode Marketing	87
2. Deal-a-Day	119
3. Facebook - Ads	157
4. Facebook - Fan Page	159
5. iPhone/Smart Phone Apps	181
6. LinkedIn - Ads	183
7. LinkedIn - Profile	185
8. Mobile Marketing	195
9. Paid to Read (PTR)	215
10. Pay per Click (PPC)	219
11. Podcasts	227
12. Proximity Advertising	247
13. Text Messaging	323
14. Twitter	331
15. Video Games - Advergaming	335
16. Video Games - In Game	337
17. Vodcasts	345
18. Voice Blasts	347
19. Website	355
20. YouTube - Ads	375
21. YouTube - Videos	377

Marketing is one of the most exhilarating industries on the planet.

Photo at right:
Taken by Charlene Brisson
in Las Vegas, 2009.

Chapter Six
-148 Profiles-

Photo: Captured from http://www.skywrite.com

References & Examples

http://www.skywrite.com

http://www.advertisingaerial.com/

http://arnoldaerial.com/marketingfacts.html

http://michiganaerialadvertising.com/

http://www.usairads.com/

http://www.nationalskyads.com

When to Use

- To geographically target large outdoor events or densely populated areas.
- Excellent for product that is available at the event being targeted.
- For low-sell, short message.

Aerial Banner

Category: Offline

Frequency: Ad Hoc

What is it

Every clear blue sky cries out for a clever distraction. Aerial banners towed by low-flying aircraft (usually plane) can make a big impact when flying over an outdoor event, crowded beach or a densely populated community. Banners are most often rectangular in shape and feature short, bold messaging for best effect. The plane travels back and forth over a specific geographic area.

Arnold Aerial Advertising quotes the statistic of 79% recall from a 2005 Miami Beach aerial advertising survey.

Pros

- Captivating, noticeable, eye-catching.
- Easy to geographically target.
- When people hear a plane most automatically look up.
- Potentially high recall rate.

Cons

- Message must be very short and memorable.
- Limited time exposure.
- Reliant on weather.
- More difficult to demographically target.

Photos: Captured from http://worldwideskyadvertising.com

References & Examples

http://www.nationalskyads.com/skywriting.html

http://www.skywrite.com

http://worldwideskyadvertising.com/Skywriting.html

When to Use

- To geographically target large outdoor events or densely populated areas.
- For low-sell, short message.
- For product that is available at the event being targeted.

Aerial Skywriting

Category: Offline

Frequency: Ad Hoc

What is it

What a terrific way to get the attention of outdoor event goers or beach visitors on a nice day. In aerial skywriting, an advertising slogan, message, or logo is spelled out by an aircraft (usually plane) that flies across the sky using a controlled emission of smoke. With this tactic the message must be short and memorable.

Pros

- Captivating, noticeable, eye-catching.
- Very unusual so people will notice.
- Most people automatically look up when they hear a plane.
- Easy to geographically target.

Cons

- Can be very costly.
- Very limited exposure.
- Only works with very short messages of 4-12 characters.
- Highly reliant on weather.
- More difficult to demographically target.

Pros

- Can significantly increase sales by having people out there selling for you.
- Increases product recognition and brand awareness.

References & Examples

http://tinyurl.com/148Ways-Robeez

http://www.sears.ca/content/corporate-info/business-opportunities/affiliate-program

http://www.tonyrobbins.com/affiliates/

https://affiliate-program.amazon.com/

Cons

- Have to do fulfillment of the product/service to the customer.
- Tracking affiliate sales and payments requires software - otherwise can be disasterous.
- Can't be certain affiliates follow same ethics or maintain like image.
- You are ultimately responsible for whatever the affiliate does.
- Potentially added customer service duties.

When to Use

- To build an army of sales people without the overhead.
- When you have a product/service others can easily sell for you.

Affiliates

Category: Offline / Online

Frequency: Regularly

What is it

Affiliates are people who help sell your product and/or service in exchange for a percentage of the sale. Some affiliates are strictly online and others are face-to-face *and* online. The best affiliates are those that have large databases of contacts. Serious online affiliates will use several online means to sell your product such as pay per click, blog postings, dedicated websites and email blasts to their database.

Many online affiliates sell groups of products and will place yours with other products on their website.

Offline affiliates use a combination of email blasts, telephone calls, promotional CD's, one-on-one meetings along with small and large events.

Screenshot: Captured from http://www.sears.ca/content/corporate-info/business-opportunities/affiliate-program

Photos: Captured from http://www.ryanair.com/en/advertise

References & Examples

http://www.onboardeuropa.com/mediacenter@2-images-6

http://www.ryanair.com/en/advertise

When to Use

- To launch a product.
- For image advertising.
- To capture specific target market on a specific airline route.

Airplane - Interior

Category: Offline

Frequency: Regularly
Adhoc

What is it

Overhead bins, the bulkhead and around the windows we go. Some airlines are now making the inside of their planes advertiser friendly. Terrific if you can brand the entire inside of the plane for one of the most captive audiences available. With in-air internet connectivity becoming ubiquitous, why not include a text or internet contest as bored passengers look for something to do.

Many airlines are catching on to this new revenue stream.

Pros

- Captive audience of frequent flyers.
- In-air internet opens opportunities to include text/email direct response component.
- Has a subconscious effect when it's in front of passengers during the entire flight.
- Could become known as the XX flight. XX=your brand.

Cons

- Not all airlines offer this advertising.
- May encounter resistance to advertising from passengers.
- May be considered "too much" if creative is too agressive and if a different advertiser owns the other inside pieces.

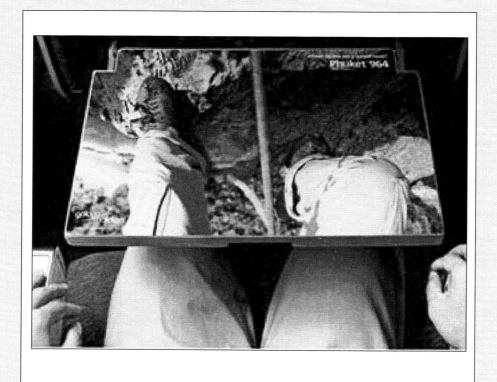

Photo: Captured from http://www.directdaily.com/?p=4533

References & Examples

http://www.brandconnections-skymedia.com/index.html

http://www.onboardmedia-group.com/

http://www.milehighmrkting.com/products.htm

When to Use

- To capture a specific target market on a specific airline route.
- Image advertising.
- For product launch.

Airplane - Tray Tables

Category: Offline

Frequency: Regularly

What is it

Remember when airlines used to feed passengers? Well that's what the seat tray was originally created for - the one that snaps up into the back of the seat in front. Now the tray's main purpose is becoming the latest "can't avoid" advertising opportunity on airlines.

Advertisers can put messages on the back of the tray so that passengers see it throughout the entire flight when it's locked into position and. The inside top of the tray is also prime real estate where the messaging can be seen when the tray is down and in use.

Pros

- Captive audience of frequent flyers.
- As internet in-air becomes ubiquitous, great opportunity to include a text/email direct response component.
- Has a subconscious effect when it's in front of passengers during flight.

Cons

- What percentage of the audience are one-trip only?
- More advertising may be a turnoff to passengers.
- Not all airlines offer this type of advertising - yet.

Photo: Taken at LAX by Charlene Brisson.

References & Examples

http://www.commarts.com/exhibit/airport-advertising.html

When to Use

- To target a traveling market.
- For image advertising.
- To launch a product.
- As part of a larger in-airport campaign.

Airport Security Buckets

Category: Offline

Frequency: Regularly

What is it

Airport security may be a serious matter, but surprisingly it isn't exempt from advertising. The bottom of security buckets are open game to get your message out.

Zappos.com has done a terrific job developing targeted messaging to a travelling market.

Pros

- Definitely a captive audience - passengers can't help but see the ad when taking things out of the bucket.
- Easy pickings if targeting a "traveling" market.

Cons

- The ad is mostly covered up by shoes, coats, bags and computers.
- People can miss picking up items because of the busy advertising.
- Many people are too concerned about catching a flight or getting through security to notice.

Photo: Taken by Charlene Brisson from televised AA Hockey Game.

Pros

- Repetition. Many ticket holders are repeat attendees.
- Usually includes an ad in each game program.
- Can often negotiate the rates.
- Get first right of refusal for the following season.
- Very inexpensive in small community venues.

Cons

- Have to pay more for playoff games.
- Can get stuck in a poor location if not purchased early enough.
- The more high profile the team, the bigger the price.

References & Examples

http://www.ddisigns.com/dasherads.htm

http://www.soccercentralindoor.com/Advertisements/advertise-with-soccer-central-indoor.html

When to Use

- To capture a dedicated sporting audience.
- To gain community credibility.
- To access a nationally televised audience of a specific demographic.

Arena Boards

Category: Offline

Frequency: Seasonal Ad Hoc

What is it

The boards at playing level around hockey, basketball, baseball, tennis, indoor soccer arenas, outdoor soccer pitches and beach volleyball play areas offer up excellent opportunities to reach sports fans.

Also called dasher boards, the size is limited so a logo is the best use of the space. Bigger is better for logo size so that even the cheap seats can get an unencumbered view. Unless your logo is weak, then consider your URL/website address.

If televised, strategic placement is wise, since television cameras are usually placed on only one side of the venue. One wants their board near the net/side of the weaker team as that's where all the action will be - but to ensure full coverage, having boards behind both sides is really the most advantageous strategy.

In the movie *Tooth Fairy* you can see an excellent example of rink boards as The Rock (Dwayne Johnson) skates around as the team's enforcer. Coca Cola likely paid dearly for this product placement.

Photo: ATM Machine taken in Richmond BC by Charlene Brisson.

References & Examples

http://www.cashngo.com/atm-machines-options.php

When to Use

- If cash machine is in the same area/location of your establishment.
- For immediate response sales.

ATM Machines

Category: Offline

Frequency: Regular

What is it

The last time I grabbed some cash from the ATM at the local shopping mall, low and behold, a promotion popped up on the screen followed up by a discount coupon that was ejected BEFORE my money and BEFORE my receipt. The discount was for merchandise in a store in that very same mall. Terrific targeting. While waiting for the money to pop out, one doesn't have to look at a blank screen any more - they can view your message.

Pros

- Captive audience with money in hand to spend.
- The coupon makes it easy to track ROI.
- Great when the deal is for a product/service that is near by.
- Coupon output is looked at because people think it's the receipt.

Cons

- The screen impression is very short.
- Can be seen as just another piece of paper adding to the landfill.

Photos: Bags made out of recycled plastics taken by Charlene Brisson.

References & Examples

http://www.customplasticbags-direct.com/index.html

http://www.icegreen.ca/

When to Use

- To replace all in-store bags.
- As a walking billboard.
- To hand out at tradeshows for people to collect items in

Bags - Eco-Friendly

Category: Offline

Frequency: Regularly

What is it

Do you use bags at your establishment so people can carry their merchandise out? Consider your bags to be walking billboards. Stay away from plastic - unless it's bio-degradable. Slap a great big logo and website on each side along with an easy-to-read eco-friendly message so everyone knows you care about the universe - something like "XX cares about my planet, this bag is bio-degradable".

Many grocery stores are selling branded reusable bags made from recycled materials. Also an excellent idea for large retailers with frequent customers or to use as a giveaway bonus for purchasing over a certain amount.

Pros

- Puts your company on the map as being environmentally conscious.
- Can use the bag to share your environmentally friendly message.
- Check-out people can reinforce the message when using them.

Cons

- More expensive than regular plastic bags.

Photos: Taken in Vancouver, BC by Charlene Brisson and Turkey by Andre Phaneuf.

References & Examples

http://www.ballooning.fsnet.co.uk/

http://www.csaballoons.com

http://www.balloonprinting.com/

When to Use

- To indicate a big sale, promotion or event.
- To launch a new product or service.
- To draw attention to your location.

Balloons

Category: Offline

Frequency: Ad Hoc

What is it

Balloons are a terrific way to bring attention to your location. For little cost you can print your logo on the balloon face and tie a bunch to a sandwich board outside your front door. From small manually and helium filled balloons right up to giant hot air balloons floating high in the sky.

Of course, the larger the balloon/s, the greater the impact - although never under estimate the impact that a bunch of inexpensive brightly colored balloons can have on an otherwise boring street or clear blue sky.

Pros

- Balloons, large or small always capture attention as they represent a "special event."
- Perfect for branding campaigns.
- Can be viewed by many, when placed/flown in high traffic areas.
- Small are very inexpensive.

Cons

- Hot Air Balloons are expensive - usually, the larger the balloon, the larger the company using it for advertising.
- Hot Air Balloons must be launched by specialists.
- Must be kept inflated.

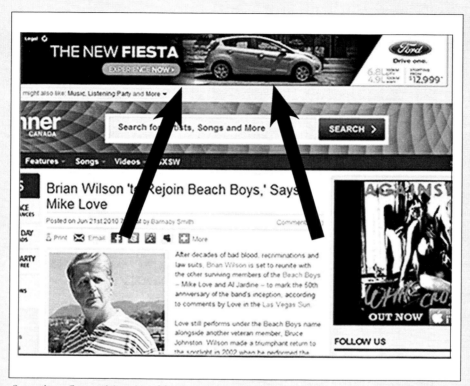

Screenshots: Captured from http://www.theweathernetwork.com/

References & Examples

http://www.submityourarticle.com

http://www.techfuels.com/general-internet-terms/1737-banner-ad.html

When to Use

- For image advertising.
- To drive traffic to your website or landing page.
- As a component of a large online campaign.
- To target geographically.

Banner Ads

Category: Online

Frequency: Frequently

What is it

Banner Ads led the revolution to monetize the internet. They're called banners because they are long rectangular in shape. The ads most often head up a web page and when clicked lead the user to the advertiser's website or landing page which provides the full details of the product, service or promotion. For best results online ads will lead to a customized landing page which will speak directly to what the ad promotes.

More and more advertisers are using dynamically changing graphic ads to attract more clicks.

Pros

- Can zero in on your target market based on website content.
- Low cost per impression.
- Dynamic banner ads, attract more clicks.
- Will bring the user directly to the page you want.
- Easy to measure ROI if analytics set up properly.

Cons

- Static ads easy to ignore.
- Debatable how effective banner ads are as they often have the lowest click thru and the lowest conversion rate.
- When buying large inventory (quantity) from publishers it can be difficult to target a specific market.

Pros

- Very simple - any photo taking smartphone and all iPhones can use.
- The barcode reader application is free.
- Can track results in realtime.
- Flexible in what you use your barcode to deliver and/or capture.
- Leverages traditional media with new media to track results.
- Different payment plans for businesses with some providers offering 30 days free.
- Non-profits can use payment technology to collect donations from printed materials, TV and on clothing.

References & Examples

http://www.scanlife.com

http://tinyurl.com/148Ways-Barcode-Ikea

http://www.mobile-barcodes.com

http://2d-code.co.uk/

Scan on your smartphone to upload Charlene Brisson's 3-Step Marketing contact data.

Text 43588 to download the ScanLife app.

Cons

- Barcode solution providers are fighting for who will reign supreme - so be cautious about who you get set up with.
- Can be confusing for the consumer - not all apps read other apps' barcodes.
- May have to educate customers and prospects on how to use the technology.

When to Use

- To capture customer/prospect data.
- To provide additional information.
- For free bonus downloads.
- As an immediate user discount bonus.
- To engage customers if target market are ubiquitous smartphone users.

Barcode Marketing

Category: Mobile

Frequency: Regularly

What is it

Wow! Barcode marketing is a fabulous way to deliver results connecting traditional media with new mobile marketing tactics. The most popular barcodes are called QR Codes.

Marketers can set up a 2D barcode through a provider like www.scanbuy.com where you can attach a mobile-enabled landing page to deliver content or free downloads such as wallpaper, videos, podcasts, etc.

Smartphone/iPhone users download the barcode reader application free of charge and use it when and where they see a barcode that entices them to engage.

The barcode can be placed on pretty much any surface available such as all types of collateral material, newspaper ads, billboards, most signage, on a TV screen, website, piece of clothing or in a video.

Barcodes are also being placed on the back of business cards. When scanned the contact details are loaded directly into the smartphone/iPhone address book along with Facebook and Twitter links. You can even program barcodes to open a pre-populated email; send a text; or connect directly to your customer service line. For events, a barcode can load the event dates into the smartphone/iPhone calendar. If you have several options for the scanner, a choice menu can be set up.

Photo: Captured from http://www.wizmark.com

References & Examples

http://www.gasstationadvertising.com/category/business-advertising/stealth-advertising/

http://www.newad.com/en/advertising2.php

http://www.wizmark.com/

http://www.inadinc.com

When to Use

- For image advertising.
- To sell alcohol or a food item that is sold in the venue.

Bathroom Ads

Category: Offline

Frequency: Ad Hoc

What is it

Inside bathrooms, specifically on the inside door of stalls, beside the mirrors and above hand dryers and urinals, are often framed ad posters. I've even encountered an ad with an embedded motion detector. It went off each time I moved into a certain position. It really scared me until I figured out what it was. In some restaurants you can find mini television screens built into the inside of the stall door. Men's bathrooms have everything from TV screens above the urinals to talking inserts at the bottom of the urinal.

This advertising is often used to promote alcohol, automobiles, media, feminine hygiene and pharmaceutical products.

Pros

- Have a captive audience - particularly inside stalls and above urinals.
- Low cost per impression.

Cons

- Limited reach.
- ROI may be difficult to measure.

Photo: Taken in New York Times Square by Charlene Brisson.

References & Examples

http://tinyurl.com/148Ways-Pattison

http://www.entrepreneur.com/advertising/adcolumnistroyhwilliams/article63692.html

http://www.metronews.ca/vancouver/local/article/535914--thieves-beware-sign-would-be-fools-gold

When to Use

- For image campaigns.
- For product launches.
- To target geographically.
- As a component to a large integrated campaign.

Billboards

Category: Offline

Frequency: Frequently

What is it

Billboards are large advertising panels, walls, or boards commonly located along high traffic roads, highways and intersections. They range in size from around 10`H x 20`W to 28`H x 60`W. The larger the size, the more expensive the lease with pressure to commit for longer time periods.

Successful ad messages are visual and kept short to create a lasting impression. Advertisers are stretching the limits as they test the boundaries in the quest for more attention. Pretty much anything goes on billboards.

Pros

- Can be very clever with the creative.
- Clever creative can produce a powerful impact.
- Are up 24/7 and most have lights so they can be seen at night.
- Provides multiple impressions to regular commuters.

Cons

- Ad message can become "invisible" once consumers become used to seeing the billboard.
- A bigger and more creative (and usually more costly) billboard is needed to stand out.
- Some billboard locations are next to impossible to get viewers.

Photo: Taken in Bellingham, WA by Charlene Brisson.

References & Examples

http://www.drobe.co.uk/article.php?id=2385

When to Use

- To launch a product.
- To point out your location.
- To target a tight geographic area that has high pedestrian and car traffic.

Billboards - Human

Category: Offline

Frequency: Frequently Adhoc

What is it

Human billboards come in many forms. By dressing up someone in an unusual costume with signage in their hands and placing them in front of your location can attract some serious attention. Most common are gorillas, but as you can see (left), the more bizarre the outfit, the greater the interest.

Sandwich boards can also be human walking billboards. Messaging is printed on two large boards that are hinged at the top and hung over a person's shoulders. The boards literally "sandwich" the person as he or she walks around designated areas, serving as human mobile advertisements. Handing out flyers while walking around is common.

Pros

- Relatively inexpensive.
- Can be attention getting.
- If seen often enough, the frequency can leave an imprint.
- A creative costume is always a winner.

Cons

- Easy to ignore if not clever enough.
- Sandwich boards can be associated with poverty and doomsayers.
- Can be so distracting to cause danger for drivers.

Screenshot: Captured at http://www.webdesignerwall.com/

References & Examples

http://wordpress.org/

https://www.blogger.com/start

http://automarker.net/blog/ping-list/

When to Use

- To build relationship with customers.
- To increase search engine rankings (SEO).
- To provide educational value.
- To show transparency to customers and prospects.

Blogging

Category: Online

Frequency: Regularly

What is it

Blogging is basically writing down all of your thoughts as they relate to your business, product and/or service. A blog is a way to build relationship with customers and prospects by sharing information that can be of help and/or inspire dialogue. It is a powerful means to get people returning to your site and to increase rankings in the search engines (SEO) by using automated distributors. Karen Bodie has posted a nice long list with instructions on her site at auto-marker.net/blog/ping-list

Free blogging platforms like Wordpress make it easy and inexpensive to set-up and get started. You can also easily attach your Wordpress blog to your existing website or use Wordpress to create an entire website for free.

Pros

- Ping lists will distribute to blogger sites and raise your search engine rankings.
- Followers will stick with an interesting blog.
- Staffers can contribute, doesn't have to be one person.
- Relatively easy to start.

Cons

- Blogging takes time and commitment - if you don't have either, don't start one.
- Some ecommerce gurus tell businesses if your blog is connected to your site to not sell on your blog - keep it information based.

Pros

- Gives credibility to your company/product/service if professionally produced and high quality.
- Can be relatively inexpensive to produce.
- Can be generic and drive to website for more info.
- Can be detailed to tweak the reader's interest to find out more.

Cons

- Don't print it if you're not going to use it.
- Schedules and dates change quickly - include disclaimers, point to website for the latest.
- Gives poor image if not professionally produced, on cheap paper, typos, one color, poor design, out-of-date information.

References & Examples

http://www.ehow.com/how_5263179_fold-paper-brochure.html

http://wiki.scribus.net/images/d/d0/BrochureFolds.jpg

http://blog.lib.umn.edu/fitzp076/architecture/brochure-folds.jpg

When to Use

- A necessity to include in face-to-face leave-behind sales packages.
- For high-end customer acquisition.
- When the image of your product/service is critical to the sale - ie resorts, spas, real estate, painters, carpenters, food, etc.
- When an educational sell is required.

Brochures

Category: Offline

Frequency: Regularly
Ad Hoc

What is it

A brochure is generally a mainstay of hard copy collateral materials that businesses use in their sales kits and to hand out to serious prospects.

Also known as pamphlet or leaflet, a brochure is any promotional piece printed on both sides that is folded. If it's not folded - it's either a rack card, a fly sheet or a flyer. A brochure can be a simple 8 1/2" x 11" one or two fold, or several pages with a staple binding. I'm looking at a piece right now for a resort - it's an 8 1/2" x 14" glossy finish folded in half and then folded again (called a double fold). Another common size is 11" x 17" folded up into an 8.5' x 5.5".

Common brochure folds are roll fold/tri-fold, double fold, accordion fold/w fold, gate fold - pretty much as many ways that you can fold a piece of paper, there's a name for it. Your printer and/or designer will work closely with you to determine what size and fold is right for you. Check out the variety of folds in the references at left.

Plain websites that aren't interactive are called "brochureware."

Photo: Taken in Vancouver, BC by Andre Phaneuf

References & Examples

http://www.elitemediainc.com/

http://www.pyramidvisuals.co.uk/services/building_wrap/

When to Use

- To make a major impact.
- With a big budget.
- To launch a new product.
- To show participation during a major event - like the Olympics, Summer Games, etc.
- To cover up renovations.

Building Wrap

Category: Offline

Frequency: Adhoc

What is it

The movie *Lost in Translation* filmed in Tokyo, shows terrific examples of an entire building wall draped in one top to bottom ad.

Building wraps are quite common in large international cities, not so much in small cities. Often draped down over 20 or more stories, this advertising tactic makes an enormous impact for your product/service. Definitely bigger than life and attracts attention.

Pros

- If positioned properly, can't be missed - can make a huge impact.
- Can show support for event.
- If blocked out of sponsored events, can place in the nearby municipality.

Cons

- Can be very expensive.
- Most often can't be used within boundaries of events unless an official sponsor.
- Need permission of building management.
- Blocks out some of the sun of the tenants in the building.

Photo: Captured from http://www.omaccanada.ca/Sites/omac/images/AboutUs/benchsm.JPG

References & Examples

http://www.benchad.com

http://www.metrobench.com/

http://www.gaebler.com/Bus-Bench-Advertising-Costs.htm

When to Use

- As a component of a large campaign.
- As a local business to target geographically.
- Good for image campaign.

Bus Benches

Category: Offline

Frequency: Frequently Adhoc

What is it

Bus bench advertising has been around for decades. Ads are usually placed on benches found in public transportation stops and areas where there's high commuter traffic.

As you can see at left, bus bench advertising is becoming very clever. Over the past few years many large municipalities have removed benches and replaced them with bus shelters - also an advertising opportunity.

Pros

- One of the least expensive forms of outdoor advertising.
- Is up 24/7.
- Can provide multiple impressions to target market since commuters regularly pass by the same route.

Cons

- Can easily be and are often vandalized since the benches are street level.
- Can be seen as eyesore.
- When buts are on the bench - your ad can't be seen.
- Larger cities are replacing the benches with shelters.

Photo: Taken in Vancouver, BC by Charlene Brisson.

References & Examples

http://www.gaebler.com/Bus-Advertising-Costs.htm

http://www.clearchanneloutdoor.com/products/bus_transit.htm

When to Use

- Great for image advertising.
- As a component of a larger integrated campaign.
- To target specific geographic markets.

Buses

Category: Offline

Frequency: Frequently Adhoc

What is it

Interior and exterior advertising can be placed on and in buses. On the exterior, advertisers can feature their message on both sides as well as the back of the bus. Some advertisers now take over the entire bus exterior which is called a wrap. Illuminated exterior night ads are also available now. They look fabulous and make a terrific impact! Interior advertising consists of messaging cards placed above the windows.

This same advertising is mostly applicable to trains, skytrains and other public transport vehicles.

Pros

- Low cost per impression.
- Can provide multiple impressions to a wide array of demographics in similar areas.
- Non-intrusive as buses are already part of a commuter's environment.
- Excellent to target Postal/Zip routes.

Cons

- If the ad is too busy, it will fail as viewers have only a few seconds to catch it.
- Repetition is critical - buses are mobile so advertisers rely on commuters' repeated exposure for their message to stick.

Screenshot: Captured at http://www.inc.com/ss/10-most-creative-business-cards#2

References & Examples

http://www.inc.com/ss/10-most-creative-business-cards#0

http://tinyurl.com/148Ways-VistaPrint

http://www.trafficdesign.ca/worldsgreenestbusinesscard.html

http://www.plastekcards.com

http://www.workthepond.com

When to Use

- At ALL networking events.
- Whenever meeting a potential prospect.
- To help customers remember your name.
- To place on the counter of your business so customers/prospects can easily take one.

Business Cards

Category: Offline

Frequency: Regularly

What is it

Business Cards are one of the primary advertising tools of all business executives. Information on your card should be basic with your contact information large enough to be read by someone over 40 at arms length. Make sure to include your value proposition (sell line) and "call to action" on the card.

Print on both sides of the card to maximize the real estate. To make your card memorable produce your cards on an odd shape so that it stands out among all the others - be creative so that your card doesn't end up in recycling.

It's funny how many people are caught networking without cards. Actually not so funny, just bad marketing.

Pros

- Simple, inexpensive means of advertising.
- Can use both sides to print your value proposition and "call to action."
- Very easy to carry around.
- Shows respect for the recipient by making it easy for them to recall your name while in your presence (and later).

Cons

- Ineffective cards with extra small print or difficult-to-read curly q fonts are frustrating for users and may be quickly discarded.
- Cards are often stuck in a pile or in a drawer and never used.
- Odd size cards may be tossed as they don't fit card holders.

Photo: Taken in Vancouver, BC by Charlene Brisson

References & Examples

http://www.clearchanneloutdoor.ca/venues/transit.htm

http://www.emcoutdoor.com/transit_shelters.htm

When to Use

- As part of a larger integrated campaign.
- To a specific geographic target market.
- For image advertising and brand awareness.

Bus Shelters

Category: Offline

Frequency: Regularly

What is it

Most major cities have replaced standard bus benches with bus shelters. Also known as transit shelters, not only do they act as shields from the elements for bus riders, they also provide a terrific opportunity for advertisers to get their message out. Large enough to be seen by pedestrians and drivers, bus shelters can be a reasonably inexpensive advertising tool to use within an integrated campaign, particularly in targeted geographic areas.

Shelter advertising can be basic or backlit to generate greater evening exposure.

Pros

- Can provide big exposure with the right creative.
- Captive audience of people waiting for the bus.
- Good drive-by exposure.
- Nice enhancement to a larger integrated campaign.

Cons

- Easy to be vandalized with graffiti.
- Can become part of the background and ineffective if up too long or the creative is boring or too busy.

Photo: Taken by Charlene Brisson.

References & Examples

http://www.catalogs.com/

http://www.gardenlist.com/

When to Use

- To present a large inventory of products.
- To reach an older "mail-order" target market.
- As a direct mail piece.
- When you have an online business and many products/services.
- To increase wallet share.

Catalogue

Category: Offline
Online

Frequency: Annual
Seasonal

What is it

Perhaps the most well known printed catalogue is put out by Victoria's Secret. Catalogues feature numerous products that a business sells, and not always by the same manufacturer. In the early days, before environmental awareness and skyrocketing delivery and postage costs, catalogues were freely mass distributed door-to-door. Now catalogues are mostly a perk after purchasing (on or offline), or as pick-ups at the store. Most companies will ship if specifically requested.

Online and printed catalogues are an excellent way to show prospects and customers the vast offerings of your business.

Pros

- Less expensive to display all items, rather than take up bricks and mortar space.
- Increases add-on sales.
- Easy to search online and easy to update.
- Long shelf life. Printed catalogues more difficult for people to toss than newspapers.

Cons

- Catalogues can be very expensive to produce - not just the printing, but photography and layout as well.
- Can be a lot of waste in distribution.
- Must keep updated as prices and inventory changes often.

Pros

- Very inexpensive.
- Can zero in on your geographic target market.
- Makes consumers feel "special," like everybody gets the product at its regular price but they're getting it at a discount.
- Good to drive people to purchase online by providing a promo code on the coupon.
- Can do testing at very little cost.

Cons

- Most get ignored/tossed in the garbage - seen as junk mail.
- Your coupon can get lost in the middle of the dozens of others in the booklet.

References & Examples

http://thatcouponbook.com/Advertising_Info.html

http://blog.nielsen.com/nielsenwire/consumer/coupon-enthusiasts-drive-up-redemption-rates/

When to Use

- When your establishment is within the community that the books are distributed.
- To encourage people to purchase within a specified deadline.
- To target geographically.

Coupon Books

Category: Offline

Frequency: Frequently Adhoc

What is it

Coupon books contain individual redeemable certificates of discounts, premium items, freebies and other promotional activities for products and services. They are usuallly redeemable by store merchants and/or online. Most often, each coupon within the book comes with a deadline, so buyers are driven to use the discount within a given promo period.

Almost all types of businesses, from supermarkets to direct-selling and service companies use this form of advertising to encourage immediate purchase. The booklets, which come in many sizes are usually distributed door-to-door as unaddressed admail or inserted into community newspapers.

Photo: *Taken by Charlene Brisson.*

Photo: Captured from www.entertainment.com

References & Examples

http://www.entertainment.com

When to Use

- To attract new customers to your establishment.
- To be seen as supporting the local charity.

Coupon Book - Entertainment

Category: Offline

Frequency: Annual

What is it

Bricks and mortar stores, services, restaurants and recreational activities tend to generate reasonable business from entertainment coupon books. People purchase these annually and then plan their entertainment for the year. The books are usually quite thick, therefore seen as great value for the price. Each page has two or three coupons and are regionally targeted.

Proceeds go to a local charity which make the books attractive for groups to sell.

Pros

- Seen as good corporate social responsibility.
- Discount or 2-for-1 brings in new customers.
- Can entice new customers.
- Relatively inexpensive to advertise in.
- Community focused as many buy the book year after year and share with friends.

Cons

- Usually one of dozens/hundreds of same industry type in the book.
- Many people buy the book to support the charity and don't use.

Photo: Taken by Charlene Brisson.

References & Examples

http://www.yourtownmailer.com/

http://www.openandsave.com

When to Use

- Commonly used by small local businesses and franchises to target geographically.
- To encourage people to purchase within a deadline.

Coupon Envelopes

Category: Offline

Frequency: Frequently

What is it

Coupon Envelopes are just that - an envelope filled with coupons and special deals. The individual front and back full color printed inserts feature a whole variety of businesses - usually local - and contain special deals on products and services. Mostly small businesses and franchises tend to use this tactic to sell carpet cleaning, roofing, house cleaning, insurance, garbage pickup, and products such as draperies, home alarms, windows, furnaces, air conditioners, and the like. The outside front and back of the envelope is also covered with offers. The envelopes are delivered as unaddressed admail through the post office. The company that puts the envelopes together handles all the delivery.

Pros

- Inexpensive means of advertising.
- Inexpensive to test.
- Can zero in on your geographic target.
- Easy to participate, the mailing company does all the mailing tasks and will design your coupon.

Cons

- High toss rate (thrown in the garbage).
- Your ad slip (coupon) can get lost with the dozens of others in the same envelope.

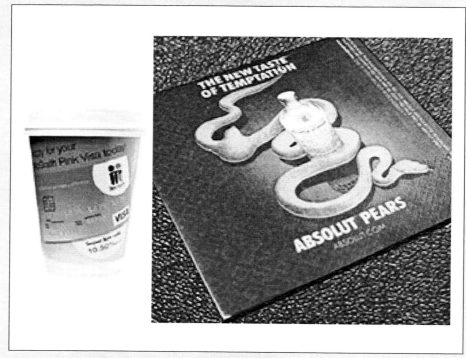

Photo: Captured from http://www.eatmedia.com.au and http://www.napads.com/napkins.html

References & Examples

http://www.britevision.com/home/index.php

http://www.eatmedia.com.au/

http://www.showyourlogo.com/custom-coffee-mug/custom-coffee-mug.htm

http://www.napads.com/napkins.html

When to Use

- Events and receptions.
- To serve coffee, tea or water in-store to your customers.

Cups & Napkins

Category: Offline

Frequency: Regularly / Adhoc

What is it

Branded coffee cups and napkins are necessary for all events that you participate in or host. Make sure they're bio-degradable and made out of recycled material. Many event organizers are looking to cut back on costs so your donation of cups and napkins are often welcome contributions to events and great branding for your copany.

Coffee cup sleeves are less expensive than the full cup and a common means of keeping people from burning their hands.

Pros

- Simple branding opportunity.
- Shows that you get it's about the details.
- Looks great and provides consistency.
- By placing branded recycling bins in the area, shows your company is environmentally conscious.

Cons

- Will cost a bit more than unbranded cups and napkins.
- Recycling doesn't always make it and ends up in the trash. Confirm where your recycling is going.

Screenshot: Captured from http://www.groupon.com/tucson/

References & Examples

http://www.groupon.com

http://deals.woot.com/sellout

https://deals.livingsocial.com/deals/how_it_works

http://www.getclip.ca/ (Canada Only)

When to Use

- To test a new product at a reduced price.
- To identify a new market.
- To target a specific geographic location.

Deal-a-Day

Category: Online, Mobile
Frequency: Frequently

What is it

There are dozens of "deal-a-day" sites on the internet. The concept is that each day one targeted product or service is posted, emailed or sent by text to signed up members. One twist on a city specific deal site is that there has to be at least three people or more that buy before the deal becomes valid. Some of these sites like http://www.deals.woot.com allow community postings which is where you can post your product deals and also pay extra to be on the high profile sponsored list of deals. All of these sites are looking for businesses to provide deals - products and services. Terms and costs are discreet as you must email to find out.

Pros

- Most service sites do geographic targeting.
- Low cost to participate.
- Can get a feel for the interest in new products or services.
- Easy to introduce your product/service to a new market.

Cons

- Product or service has to be seen as a really good value or don't bother.
- Many sites only work with products, not services.
- Easy to geographically target, difficult to demographically target.

Photos: Captured at http://www.toxel.com/inspiration/2010/03/02/clever-uses-of-stickers-in-advertising/

References & Examples

http://www.toxel.com/inspiration/2010/03/02/clever-uses-of-stickers-in-advertising/

http://www.directdaily.com/?p=3445

http://www.directdaily.com/?p=7073

When to Use

- To make a major impact in an unusual way.
- For a temporary campaign.
- When requiring only a small quantity.
- To test a campaign in a small area before expanding into a large print job.

Decals

Category: Offline

Frequency: Adhoc

What is it

Do bumper stickers come to mind when thinking of advertising decals, also known as stickers? Undoubtedly, they are considered as part of this tactic, but decal creative and size have come a long way from the back end of a car. Decals can and are now being placed on any surface from ceilings, walls, windows, doors, cars, floors, sidewalks, roads and even eggs. Pretty much any surface can be stickered to get your message out. And as always, the more clever you are, the more you will attract your customer.

Pros

- Can easily produce small quantities.
- Can make a big impact with clever creative.
- Unlimited creative opportunities.
- Can place decals onto unusual surfaces to catch customers/prospects off guard.

Cons

- Can often see that the creative is a sticker.
- Can be pricey to print small quantities.

Photos: Taken by Charlene Brisson in Vancouver, BC from floor at Trade & Convention Centre.

References & Examples

http://www.power-graphics.com/pages/floor_decals.html

http://www.3m.com/intl/BE/english/front/full_03.html

When to Use

- For point-of-sale advertising - if your product or service is available in the establishment where the sticker lays.
- For on-site suggestive selling.
- As directional signage.
- For product launch.

Decals - Floor

Category: Offline

Frequency: Regularly
Adhoc

What is it

Decals can be placed on pretty much any surface and floors are one of the prime spaces being used to catch the attention of customers and prospects. Used most commonly for general messaging, floor decals are also a terrific way to direct participants through events and to lead buyers to your booth or store location in a mall.

Pros

- Can be very impactful with the right creative.
- Floor advertising is fairly new so attracts attention.
- Low cost per impression.
- Can match up with merchandising displays.
- A reasonably captive audience.

Cons

- Limited impressions per sticker.
- Easy to ignore.
- Must be very durable to avoid quick wear and tear and regular replacement.

Pros

- With addressed direct mail you can narrowly target your customer.
- Works as a component of an integrated campaign.
- Works well when sending to existing or past customers, especially if personalized or customized with a handwritten note (only doable for small quantities).
- Technological advancements can personalize your piece in several areas - the more personalized the piece is, the higher the open rate.
- Inexpensive to send unaddressed.

References & Examples

http://www.direct-axis.net/direct.html

http://www.directcreative.com/samples.html

http://direct-mailer.com

Cons

- High toss rate (thrown in the garbage).
- Name typos on addressed admail turns potential and existing customers off.
- Garbage in - garbage out. Lists are only as good as the data input - keep your customer and prospect lists clean and current.
- The older the list, the more dead mail (moved or changed jobs, etc.).

When to Use

- To warm up prospects before doing a follow-up telephone campaign (it will significantly improve response).
- When you have a clean list of very targeted customers.
- To target geographically.
- As a follow-up offer to existing customers.

Direct Mail

Category: Offline

Frequency: Regularly
Ad Hoc

What is it

Many refer to direct mail as "junk mail." This is because the piece they received wasn't targeted. If you receive a letter in an envelope to your home with your name on it, you rightfully expect that it is for you or at the very least be about something you are interested in. It's the random marketers that have been sending out billions of pieces each year to people who have no interest and never will that have given direct mail a bad name - not unlike email spam. A typical direct mail package includes an envelope, letter, a direct response sales flysheet or postcard and a self mailing return envelope. The package can be addressed to a specific person (known as addressed) or labelled as homeowner, resident or occupant (known as unaddressed). Flysheets, postcards and brochures can also be delivered as unaddressed direct mail or even stand-alone inserts that would commonly go into direct mail packages. As long as the item is delivered by the post office, it is considered to be direct mail.

The post office charges different prices for addressed vs unaddressed as well as for mail packages vs other forms. To get a reduced bulk rate for direct mail packages, each MUST contain the exact same components - including identical wording.

Photo: Unaddressed Direct Mail taken by Charlene Brisson.

Direct Mail Con't

Category: Offline

Frequency: Regularly
Adhoc

Continued from Page 125

The actual delivery is usually done within 1-4 days of dropping your campaign pieces at the Post Office.

Unaddressed campaigns can be targeted geographically down to postal walks using FSAs which are the first 3 digits of postal codes in Canada or Zip Codes in the USA. Mailers can cross reference census information such as age, household income (HHI), single family or rental dwellings, number of children, etc. to identify postal walks/neighborhoods of people who are within your target market.

Lists used for addressed campaigns can come from your own database as well as very targeted lists you can rent from list brokers. Brokers market names such as magazine subscriber lists (paid current subscriber lists are more responsive than recipients of free publications) as well as membership and general industry directories. Brokers can refine your "selects" by job title, geography, and often by behavior such as "have purchased a specific item" or "attended a specific type of event". This helps you target right in on people who are closest to your prospect.

Photo: *Addressed Direct Mail taken by Charlene Brisson.*

Direct Mail Con't

Category: Offline

Frequency: Regularly
Adhoc

Continued from Page 127

List rentals usually start at around $.25 per name. The price goes up as the effectiveness of the list increases. For example, a list of lapsed subscribers of a women's magazine is less costly than an active subscriber's list. A paid subscriber list is more expensive than a free subscriber list. You can also rent phone numbers along with the addresses. It's been my direct experience that when an addressed direct mailing is followed up with a live sales telephone call, your response rate can as much as triple. Again - only if you have a recent targeted list.

Most lists are rented which means you can only use the name once. Brokers will electronically send the list directly to your mailing house to be output directly onto envelopes or labels and then lettershopped (stuffed, postal walk sorted and dropped off at the post office). Any and all responses that you receive are yours to remail and/or telephone as often as you'd like. If you rent phone numbers and/or email addresses, commonly you only have one time use. Brokers and list originators place "seeds" in their lists so they'll catch you if you mail, call or email more than once.

Addressed direct mail usually receives a higher response overall than unaddressed.

Pros

- Listings are free.
- The directories promote and market themselves online - you don't.
- Easy to submit listings.
- Can help to increase search engine rankings.
- Can often include your specific keywords.

Cons

- Some directory owners use their listings to draw people in to click on other pay per click ads - can be an integrity issue.
- Can spend a whole lot of time submitting listings for little to no results.

References & Examples

http://www.Google.com/LocalBusinessCentre

http://ca.dir.yahoo.com/Business_and_Economy/Directories/

When to Use

- For industry credibility.
- To put yourself in front of people searching on the internet for products and services.
- To help increase your search engine rankings.

Directories

Category: Online

Frequency: Regularly

What is it

The internet is crammed with business directories where you can list your business under specific categories. Online directories differ from the Yellow Pages in that there is no printed book and their sole purpose is to drive prospects to your website. Many of the Yellow Pages and 411 online directories still don't have website listings.

Google, Yahoo and other prominent search engines are good places to make sure your business is listed. You can also be listed in Google Maps to bolster exposure. Other directories are created by internet marketers that collect listings and surround them with pay-per-click ads to generate revenue.

It's easy to get a free listing in directories by simply filling out and submitting the online application form which you can usually link to off the main page.

To find directories to be listed in, search for Business Directory and/or Business Directory Your Product/Service. Don't waste your time on listing with directories that have few existing listings.

Photos: Taken by Charlene Brisson.

References & Examples

http://www.doorhangers.com/door-hangers-templates.shtml

http://tinyurl.com/148Ways-DoorKnobAds

When to Use

- To launch a product.
- To reach a very specific household geographic target.
- To put a product (sample) right into the hands of your prospect.
- As an incentive (with coupon) to try your product or service.

Door Hangers

Category: Offline

Frequency: AdHoc

What is it

A door hanger is an advertising message that hooks onto a door knob. As a door-to-door advertising tool, door hangers can be very creative by containing sample products and incorporating coupons along with "can't miss" branding. Because the recipient has to touch the piece in order to remove it from the door knob, it has a higher likelihood of being seen than typical direct mail and flyers.

Pros

- The recipient can't miss a hanger.
- Flexible in design - and can include product samples.
- An unusual message delivery tool - so is noticed.
- Can order online.

Cons

- May not be the decision maker that takes the hanger off the door knob.
- May have to pay extra for a die cut.
- Need to monitor delivery so they're not just stuck in mailbox.

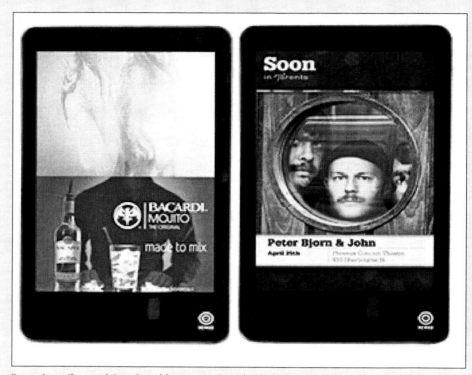

Screenshots: Captured from http://www.newad.com/en/advertising_eboard.php

References & Examples

http://www.newad.com/en/advertising_eboard.php

http://www.newad.com/en/pdf/FEATURE_SHEET_eBoards%20March09_ENG.pdf

When to Use

- For image advertising.
- When you are selling the product advertised in the establishment where the eBoard is placed.

eBoards

Category: Offline

Frequency: Frequently

What is it

eBoards look not that much different than a supersized iPhone that hangs on the wall. Dynamic advertising is mixed with nightlife clips to attract a young audience. Clever ads can be done by splitting the screen horizontally so that two different creatives are presented for the same product/service - or the entire screen can be used. The screen size is 15" or 19".

These are really cool. You have to see the demo to get the true perspective.

Pros

- Very targeted to an 18-34 demographic.
- The creative is very unique and eye-catching using lifestyle images.

Cons

- Only seen in Canada.
- Difficult to measure ROI.

Pros

- Auto-responders are working while you're not.
- Inexpensive to use service providers (based on size of database).
- Gives consistency to communication.
- Relatively easy to set up for beginners.
- Can run several campaigns at once.
- Email effectiveness testing can be done easily.

References & Examples

http://www.getresponse.com

http://www.verticalresponse.com

http://www.silverpop.com

Cons

- Need to keep testing to make sure you're getting the best possible results.
- Too many emails too close together turns people off.
- All the "cons" of email - blasts.

When to Use

- As a list builder - to capture names in return for giving something free.
- To sell a prospect after they've opted in for a free report, newsletter or video.
- To deliver an information email series such as tips, secrets, quotes, etc.

Email Auto-Responders

Category: Onine

Frequency: Regularly

What is it

Email auto-responders are automated emails that are set up in advance and programmed to be sent at designated intervals based on specific triggers. For example, if you have a sign-up form on your website - when the visitor fills it out and clicks submit, you can set-up an automated email to go out immediately to confirm their request. As soon as the recipient clicks confirm, it triggers an email to automatically go out as a thank you along with the download link to whatever they signed up for (if that's what you set the email to do and say). One day, 3 days and 7 days later (or whatever interval you choose), other emails go out to that same person with additional general, promotional and/or sales information. Once set up, the emails go out automatically all based on the initial trigger.

Triggers can be set up for pretty much any email. If they click this, then they'll receive that, etc., etc. Service providers like Vertical Response and Get Response make setting up auto-responder emails relatively easy.

Auto-responders are what you use to work while you sleep.

Pros

- Can gain greater share of wallet from existing customers by offering other products/services.
- Easy to update database about the latest promotions, sales, events product/service development, etc.
- Very inexpensive to send bulk emails as most service providers charge by size of database.
- Can reach tens of thousands of people in one blast.
- Service providers offer easy results tracking of click thrus and a simple means to manage your database.

Cons

- Send too many emails and people will unsubscribe or not read.
- Without a compelling subject line readership will suffer.
- Easy to ignore or delete.
- Email can go into junk mail if receiver doesn't add your address to their outlook/email account, also called whitelisting.
- Relatively easy to get blocked by the larger ISP's by not following best practices.
- No opt-in process makes you susceptible to being blocked by ISPs.

References & Examples

http://codex.wordpress.org/Spam_Words

http://www.getresponse.com

http://www.verticalresponse.com

http://tinyurl.com/148Ways-Forbes

When to Use

- To upsell existing customers.
- To sell prospects.
- To communicate and regularly engage customers.
- To announce a last minute sale.

Email Blasts

Category: Online

Frequency: Regularly

What is it

Email blasts are a simple and time effective way to deliver messages to large groups of people for very little cost. Emails are often designed in simple text AND html which can include branded headers, logos, photos and video links to add dimension and interactivity.

There are dozens of service providers like Vertical Response or Get Response which provide email building software along with the means to manage your customer and prospect opt-in lists and to set up auto-responder emails. Spamming, which refers to sending out bulk emails to people who haven't asked for them is not only completely unacceptable, these emails are more and more difficult to get delivered because of firewall settings.

Opt-in email campaigns give recipients an option to subscribe to a company's mailing list - usually through a newsletter or free report request - while signalling to internet service providers (ISPs) like AOL, Hotmail, Yahoo, etc. that your messages are not spam. Once you have an opt-in name, you are free to send subscribers email updates, sales pitches, newsletters or any number of blasts. Of course, you can still be blocked if using any one of a whole list of certain keywords like FREE, Buy Now, Get Rich Quick, Shipping!, Today, Here and a host of others (see reference link).

Photos: Nascar racer captured from http://www.nascar.com; mogo skateboard taken by Charlene Brisson.

References & Examples

http://www.nascar.com/

http://www.danicaracing.com

http://tinyurl.com/148Ways-Nascar

http://tinyurl.com/148Ways-Skiers

When to Use

- To build credibility within an industry.
- To get massive exposure on high profile person, usually sports figure of some kind.
- To outfit staff and street teams.

Embedded Logos

Category: Offline

Frequency: Regularly
Ad Hoc

What is it

Embedded logos are EVERYWHERE. On t-shirts, hats, jackets, sunglasses, on the bums of sweatpants and even embossed in the bottom of flipflops that leave logos behind in the sand.

To see how far embedding goes, simply watch a downhill skier or Nascar driver interviewed after a race. Logos are on their goggles, helmet, turtleneck, jacket shoulder, front, arms, on the back of the skis (usually held up to their face), on the top of their gloves and on their knees - and usually all different brands. I love watching Nascar races and seeing the logos all over the exterior and interior of the cars. Every inch of their person and equipment exposed to cameras from any angle is covered in a logo - that's embedded.

Pros

- The more high profile the person and the more high profile the placement, the greater the exposure.
- Unlimited options of where to embed logos on people and products.
- Can make an impact with single logo'd items.

Cons

- Can be difficult to measure specific ROI.
- Have to rely on the integrity of the individual wearing your logo (ie. Tiger Woods).
- Often have to share space with several other brands.

Photos: Captured at http://gearcrave.com/2008-11-19/fruitful-technique-sun-tanned-apples-with-apple-logo and http://www.totallychocolate.com/CorporateGifts.aspx

Pros

- Very unusual so it gets attention.
- Generates a lot of talk by attendees.
- Inexpensive for large quantities.
- There are a lot of different choices of candies and chocolates.

Cons

- Most items have to be used up quickly.
- Not all chocolate/candies are created equal - get samples before committing.
- Limited producers for fruit and vegetable logos.

References & Examples

http://www.gearcrave.com/2008-11-19/fruitful-technique-sun-tanned-apples-with-apple-logo/

http://www.showyourlogo.com/Promotional-Candy.htm

http://www.totallychocolate.com/CorporateGifts.aspx

http://tinyurl.com/6xaw3r

When to Use

- For special events.
- When items can be used up quickly (conferences, galas, outdoor events).
- On the counter, event check-in or reception desk.

Embedded Logos - Food

Category: Offline

Frequency: Ad Hoc

What is it

Not only can fruits and vegetables be grown into an unusual shape, but they can also be logo'd. A logo-shaped sticker (just like a tanning sticker) is placed over the fruit or vegetable just before it ripens resulting in a pre-ripened colored logo while the rest of the fruit/vegetable develops into it's rich natural color.

Another way to embed a logo on food is to brand it - like branding burgers, weiners or buns at an outdoor BBQ event. Toasters can also be customized to toast your logo into every piece of bread. These techniques are lots of fun that encourages lots of talk.

Chocolates are also a terrific way to get your brand into the hands (or mouths) of customers. These are my personal favorite. I've worked with more than one company where we used logo embossed mini bars to treat customers, prospects and event attendees. They're very popular. You can also custom brand M&M's, mints and mint boxes - pretty much any kind of candy you want. People love these!

Screenshot: Captured from http://www.epostcardxpress.com/index.html

References & Examples

http://www.epostcardxpress.com

When to Use

- When your establishment caters to tourists.//
- Trade shows and/or exhibits that cater to a global audience.
- As an event activity.

ePostcard Xpress

Category: Offline
Online

Frequency: Regularly

What is it

The e-Postcard Xpress is a large self-standing video kiosk embedded with a video camera. Customers pay a nominal amount by credit card to videotape themselves in a "live" postcard message and select one of several backgrounds. They then email the virtual postcard directly from the kiosk and/or select a hard copy print-out for themselves. Your company brand and link are included in the postcard so recipients can explore your business.

These virtual postcards are a lot of fun to record, send and receive.

Pros

- Your business gets in front of new potential prospects worldwide.
- Capture both the receiver and sender's email for future campaigns
- Revenue generator.
- Low cost for users.
- Fun and easy to use.

Cons

- Location, location, location - has to be in high foot traffic area of tourists.
- Subject to the occasional server being down or internet offline.
- Requires a fairly large floor space.

Photo: Captured from http://www.ehc-global.com/en/handrail-advertising

References & Examples

http://www.aapglobal.com/aaprails/aaprails.pdf

http://www.ehc-global.com/en/handrail-advertising/

http://tinyurl.com/148Ways-EscalatorRail

When to Use

- For image advertising.
- When product is sold or in mall/building where escalator is.
- For contest entry messaging.
- For short repetitive messaging and logo recognition.
- Event or business launch /grand opening.

Escalator Handrails

Category: Offline

Frequency: Regularly
Adhoc

What is it

Ever look down at the handrail while riding an escalator? You will if it's covered with advertising! Very clever. Works best if the messaging is short and repetitive all the way along since one usually occupies only a step or two. Average escalator ride is 30 seconds - enough time to get attention and even text into a contest.

Pros

- Is a new means of advertising in most areas - will catch big attention.
- Films cover the handrail so there is no damage to the railing.
- Can integrate with texting contests, entry codes on the railings, etc.
- Embed code/URL in creative to browse real time.

Cons

- Can be difficult to measure ROI without a direct response component.
- Graffiti could be a problem.
- Not all facilities will accept this kind of advertising - your job is to CONVINCE THEM.

Photo: Captured at http://www.directdaily.com/?p=4127

References & Examples

http://vimeo.com/2976497

When to Use

- For product launch.
- When product is sold or establishment is in mall/building where escalator is also located.
- For image advertising.

Escalator Steps

Category: Offline

Frequency: Regularly
Adhoc

What is it

Stand at the bottom of an escalator and watch each stair rise. The front of each of those steps is prime advertising real estate. Burmashave ads work really well with this method - that's when each step contains a different word or image that makes up the total message.

Pros

- Captive audience.
- Can be very clever with the messaging.
- Unusual in most parts of the world so it will attract attention.
- Perfect for short repetitive messaging and logo recognition.

Cons

- Do people really see it when on the escalator?
- Can't see if escalators are consistently crowded.
- Can be difficult to measure ROI.
- Not all facilities will accept this kind of advertising - your job is to CONVINCE THEM.

Pros

- With some simple Google research, you can find some very targeted ezines to advertise in - start by entering your topic + association + geographic area (if relevant).
- Ads are usually very reasonably priced.
- Someone else does all the work of building the list that you can access.
- Easy to measure response by click thrus (be sure to set up Google analytics or other tracker).
- Simple to set up a landing page speaking directly to the readers.

Cons

- Lack of research can result in your ad on an inappropriate ezine - know what you're advertising in, read a few past issues.
- A bad ad is still a bad ad - make sure you follow best practices to build your ads.
- Free ezines have a low readership rate - no matter how many people subscribe to it.

References & Examples

http://www.marketingexperiments.com/email-marketing-strategy/ezine-advertising-tested.html#1b

http://www.adlandpro.com/advertisers/sponsored_ads.aspx?TB=ad

When to Use

- To stay in front of a very targeted customer.
- As a complement to your article also appearing in the ezine.
- To gain entry and recognition into your industry.

Ezine - Ads

Category: Online

Frequency: Regularly

What is it

Ezines are simply online newsletters or magazines. Some are actually in the same format as a hard copy publication in pdf format and others appear in an online "newsletter" format of being relatively short and simple with the first paragraph showing up on page one and a click thru to get to the rest of the article. Just like traditional offline magazines and newsletters, each is targeted to a particular subject and subsequently there are as many ezines as there are topics one can think of.

A terrific source of targeted ezines can be found through industry member associations - almost all have a monthly electronic newsletter or magazine. Many ezine publishers are soliciting for and will accept paid advertising such as:

- display advertising;
- banner ads;
- text ads (often classifieds); and
- solo ads sent to the ezine list (see solo ads listing).

Pros

- One of, if not the best way to build targeted lists for sales conversion.
- Articles can be submitted free of charge to submission services.
- Opportunity to provide more information than a quick ad can provide.
- Editorial is always seen as more credible than paid ads - can use low sell/no sell approach to garner interest and secure info with a simple "for a free report click here" at end.
- Easy exposure for you, your company or product.
- Member organizations you belong to welcome articles from members.

Cons

- Never really sure how many people are actually reading the free ezine/newsletter that the article is posted in - much higher readership for paid ezine subscriptions.
- When using free distribution service, you can't control who runs your article.
- If your article is too salesy, it won't be read and probably won't be acceptable for content - keep it product/service information/educational based.

References & Examples

http://www.articlemarketer.com

http://www.submityourarticle.com

http://ezinearticles.com/

http://new-list.com/

When to Use

- To reach very targeted online customer segments.
- To establish credibility within your industry.
- To create online brand awareness.

Ezine - Article Marketinwg

Category: Online

Frequency: Regularly

What is it

Internet sales guru Armand Morin claims that "article marketing is one of the best forms of advertising - period. It beats almost anything on the net – because when people are online they search for articles, they don't just search for videos or twitter posts, they search for articles."

Ezine publishers are always looking for articles to fill up the pages of their online publications. Few actually pay for content. In exchange for an article, writers include their contact information (email address and website) and a line such as "click here for your free report on the above topic." Make sure that the free report you give away is about the same topic your article is about. Making the landing page an opt-in will increase sales conversion as it will be easier to capture visitor info before trying to sell them.

To get your articles posted in thousands of ezines use a submission service. Revise your article if you're using more than one service. Also search out a few paid industry association membership ezines to run your article, most will welcome it.

Ezine subscriptions and sign-ups are either paid or free. Paid subscribers ALWAYS READ MORE - so your click thru rate will be higher.

Pros

- Significantly cheaper than printed publication.
- The subscriber list is excellent lead generation of targeted prospects.
- Can use to publish other ezine publisher's articles as trade-offs.
- Can feature suppliers and/or customers articles.
- Can be revenue generating (see ezines - ads).
- Can easily create through service providers like GetResponse.com, etc.

Cons

- Pdf newsletters are often blocked by firewalls.
- Best to embed your format in the email or click through to your site.
- Must be consistent so deliver what and when you promise.
- Time consuming to create.
- If too infrequent, subscribers forget about you.

References & Examples

http://www.articlemarketer.

http://getresponse.com

http://constantcontact.com

When to Use

- To communicate to customers and prospects.
- To educate about products/services.
- To access big list owners' email for you.
- As a lead generator.

Ezine - Publisher

Category: Online

Frequency: Monthly
Weekly

What is it

Creating your own ezine will give you editorial freedom to discuss whatever it is that you find of interest to your targeted readers. The online publication can be a short one-pager all the way up to several formatted pages. In my opinion, to be considered an ezine, there must be at least three or more articles - all related to your business or product line.

Ezines are an excellent way to feature suppliers or swap articles with big list owners who will email out for you. When you run out of content - use one of the article submission sites.

If you keep the sales copy to a minimum, ezines are a great place to present a regular special offer unique to subscribers so they will look forward to each issue. If the offer is from a third party you can take 20-35% commission from all the sales. Tracking is very important if you take this route.

Pros

- Relatively inexpensive - can start with as little as $5 daily limit.
- Can zero in on your target market.
- Clickability makes it easy to drive significant traffic to your site/landing page.
- Easy to test and continually improve results.

Cons

- Very easy to ignore.
- Sales conversion decreases if landing page doesn't speak to the ad promise.
- Not all products/services have same conversion results.
- Some unethical ads that lead to regular monthly charges create user resistance.

References & Examples

http://www.facebook.com/FacebookAds

http://www.fastcompany.com/1584920/facebook-now-more-popular-than-google-let-the-ad-wars-begin?partner=rss

http://www.facebook.com/press/info.php?statistics

When to Use

- To target a specific demographic and geographic audience.
- To drive traffic to your site.
- To increase brand exposure.
- To sell online product/services.

Facebook - Ads

Category: Online

Frequency: Regularly
Adhoc

What is it

At this writing, Facebook (FB) claims to have 500 million users - 50% that log on daily. That's an enormous audience. Their demographic is wide in appeal with the largest growing user being female over 55.

Like most social media sites, you can run ads on FB. They can be seen on the right hand side of your facebook profile page. The ads are simple with picture on top and text below and includes a link. For as little as $2 per day you can run an ad that includes a small photo, link, 25 character bolded headline and up to 135 characters of body copy. Ads can be targeted so that they only appear on profiles that match your demographic and geographic choices.

Ads are bid on for viewers and you determine how much you will pay per thousand impressions (CPM) or per click thru. You have to watch your bids carefully to make sure that your ads are being seen. If your bids are too low they are simply pushed out of the way for the higher bidder and won't show up on anyone's profile. If the bid is too high, you're being seen, but wasting money - so it's important to keep watching.

A good resource is to join the Facebook Ads fan page.

Pros

- Terrific way to access some of the 400+ million people on FB.
- Can increase SEO.
- Access to people and their networks that is unavailable in other ways.
- Builds relationship with customers and prospects.
- Can be used for customer service issues.
- Provides medium to hear what people are saying and thinking about your brand - that can be acted on.
- Can post on your fan page remotely through mobile, iPhone and email.
- New applications are being launched regularly to enhance the usability.
- Can create a custom URL.

When to Use

- To establish a professional FB presence.
- To build brand credibility.
- As customer service tool.

Cons

- It takes consistency and time to keep an updated page - don't start if you can't keep up.
- Be prepared to read things about your brand that will be difficult, but necessary to hear.
- Privacy issues have created user resistance to Facebook.

References & Examples

http://www.facebook.com/help/#/help/?faq=12814

http://www.facebook.com/help/#/help/?search=fan%20pages

http://www.whitepapersource.com/marketing/facebook-landing-pages-another-white-paper-registration-tool/

http://www.facebook.com/Starbucks

http://www.facebook.com/Papajohns

http://www.facebook.com/Sprinkles

http://tinyurl.com/CharleneBrisson

Facebook - Fan Page

Category: Online, Mobile

Frequency: Regularly

What is it

Facebook has three different page types - a personal profile page, a fan page and a group page. The friend page is what most people start out with as individuals. A friend page is identified by a first and last name; has limits on the number of friends you can accept (5,000) and the way you communicate. Whereas a fanpage can be branded professionally as a business or expert; can accept limitless fans (likes); has a built-in mass communication tool (updates) with the ability to target to specific demographics; is administerable by many if you choose; and can be advertised throughout the facebook network (see facebook ads).

On your facebook pages you can also create a customized welcome landing page to greet visitors and have them sign-up for your "free product/newsletter/etc." Great list building opportunity!

Being on Facebook helps companies (large and small) to build brand credibility and relationship with customers and prospects by being transparent in providing information, running contests, answering questions and conducting customer service. A Facebook fan page is an informal way to engage your audience.

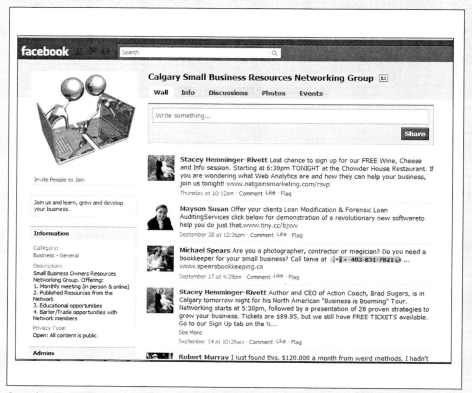

Screenshot: Captured at http://www.facebook.com/?sk=2361831622#!/group.php?gid=115868311770293

References & Examples

http://www.facebook.com/help/#/help.php?page=826

http://www.facebook.com/groups/create.php

When to Use

- To join groups that relate to your business and post answers to questions, invite members to your page and write low-sell posts about your services on the group wall.
- To attract people interested in your product focus.

Facebook - Group

Category: Online

Frequency: Regularly

What is it

Creating a Facebook Group is an opportunity to attract members that are interested in a specific educational topic relevant to your business/product or service. It's up to the group administrator to provide content-rich posts to encourage discussion. Although Facebook promotes fan pages as providing the ability for businesses and public figures to broadcast to their fans, a group page provides a much better application to actually do so. Messages sent to group members go directly into their facebook message box which in my experience, users pay more attention to. The downside can be that you'll have to message ALL members - no segmenting.

Joining other topic related groups is a great way to weigh in on issues and draw targeted people over to your fan page or drive to your website.

Pros

- You can send direct messages to group members.
- There are thousands of groups on facebook - so there must be several relevant to your business.
- A group can be set up as private or even secret (possibly for VIP clients).

Cons

- Can easily get distracted while on Facebook - set time limitations.
- Can spend all your time running a group and not achieving goals - can be better to join others and comment on discussions.
- Posts without a purpose are a waste of time.

Pros

- Relatively inexpensive.
- Can reduce distribution costs by piggybacking off others' mailings and through inserts into newspapers.
- Easy to target geographically through newspapers.
- Can print on recycled paper ro deflect environmental concerns.

Cons

- The majority are not read and thrown into the garbage.
- They have a bad rap - can be seen as environmentally irresponsible in this "green" age.
- Perceived as a nuisance if not targeted - but then isn't ALL poorly targeted advertising.

References & Examples

http://www.usps.com/send/waystosendmail/senditwithintheus/standardmail.htm

http://www.canadapost.ca/tools/pg/customerguides/CGu-naadm-e.asp

http://www.ticketmaster.com/media/directmail.html

http://www.marketingcharts.com/print/ad-inserts-capture-consumers-attention-3420/

When to Use

- For items where price point is really competitive like consumables, sporting goods, DIY, gardening, cosmetics, etc.
- To piggyback onto a mailing - yours or others.
- To promote events - sales and otherwise.
- To target geographically.

Flyers

Category: Offline

Frequency: Regularly
Adhoc

What is it

A flyer is similar to a poster, but typically smaller and often printed on lighter paper stock. The definition from http://define.com describes a flyer as "an advertisement ... intended for wide distribution." Usually, but not always smaller than 8 1/2"x 11", the word flyer is also used as a catch-all term to describe printed advertising materials smaller than a poster; collateral pieces inserted into already existing direct mailings (also known as a flysheet or handbill); and to define the dozens of sales inserts put into daily and community newspapers.

Other terms used for flyers are brochures, handbills, leaflets, circulars, flysheet.

Screenshots: Captured at http://www.ticketmaster.com/media/directmail.html

Photo: Captured at http://www.projectionadvertising.co.uk

Pros

- Can be extremely impactful in a low light venue and/or entrance.
- The creative doesn't have to be only a logo.
- Can use brand colors.
- Can use in relatively bright rooms.
- Can be very eye-catching.

Cons

- Limited impressions.
- Could be easy to ignore.
- Full color Gobos can be expensive and take longer to produce than basic Gobos.

References & Examples

http://www.gobosource.com/Custom_Gobos/custom_gobos.html

http://www.projectionadvertising.co.uk/event-branding.aspx

When to Use

- To make a branded impact at events and trade shows.
- As a directional meet-up point.

Gobos

Category: Offline

Frequency: Adhoc

What is it

Basically, a Gobo (Goes Before Optics) is a metal plate of sorts that has a logo or image cut-out and is placed over the top of a light source (Gobo projector). The cut-out is the only thing that shows through onto whatever surface it is pointed at. Full color Gobos are made out of glass and are more expensive - but also much more attractive.

Event producers use Gobos on entrance floors, walls, the back of the stage and on outdoor surfaces - pretty much on any flat surface that has audience exposure. The movement of Gobos is determined by the versatility of the light source. I've seen and used Gobos that rotate around the room, within a circle on the floor or simply in one high-profile spot.

Photo: Captured at http://crushable.com/entertainment/dominos-greengraffiti-campaign-plus-2-15-gift-cards-to-give-away/

References & Examples

http://crushable.com/entertainment/dominos-greengraffiti-campaign-plus-2-15-gift-cards-to-give-away/

http://www.dirtystreetadvertising.com/

When to Use

- As the entrance to your establishment.
- As a contest component.
- Randomly in targeted areas with lots of foot traffic.

Graffiti - Green

Category: Offline

Frequency: Seasonal
Adhoc

What is it

Brilliant way to advertise a simple logo or design. Just lay a heavy stencil on the sidewalk and powerwash the stencil. The dirt is washed away leaving your design. This is so easy and so cool. LOVE IT!

Domino's ran a two week contest in New York, L.A. and Philadelphia. The first 250 people who emailed in a photo of the green logo won a $15 Dominos certificate.

Pros

- Very clever, will attract attention.
- Lots of opportunities to integrate other activities and contests.
- Totally environmentally friendly.

Cons

- Seasonal in most locations with winter weather.
- May not last long if in well used pathway.
- Can upset other establishments when randomly placed.
- Eventually the city will catch on and insist on permits - but in the meantime, go wild!

Pros

- Keeps attendees' interest throughout the event.
- Short term event with long lasting effect.
- Can get very creative interpretations of your product/brand.
- Can establish rules of drawings.
- Could result in creative for a future campaign.
- Will likely draw media attention.

Cons

- May be difficult to control artists.
- Takes a lot of coordination.
- Requires large venue.

References & Examples

http://www.medialifemagazine.com/news2005/aug05/Aug22/1_mon/news5monday.html

http://www.altterrain.com/Nesquik_Graffiti_Performance.htm

When to Use

- At events - indoor or outdoor.
- To gather community involvement.
- To create a new and edgy graphic look for a campaign.

Graffiti - Live

Category: Offline

Frequency: Adhoc

What is it

Hire graffiti artist(s) to create a mural incorporating your brand/product onto a large, maybe even a moveable canvas. The mural or murals are created live throughout the entire event so that attendees can watch. Why not have three or four artists compete. Have attendees vote on their favorite? The only rule is that artists have to include their version of your logo or image - whatever you choose. Even point a live webcam at the canvases so that people unable to attend your event can watch the art being created online - also great for media to monitor the progress. Auction off the art after the event for charity or use it in an online and traditional ad campaign. Include a canvas that audience members can freely add their own graffiti.

Screenshot: Captured at http://www.altterrain.com/Nesquik_Graffiti_Performance.htm

Photos: Captured at http://bedstuybanana.blogspot.com/2008/04/graffiti-ads.html

References & Examples

http://www.altterrain.com/graffiti_advertising.htm

http://www.marketingminefield.co.uk/unusual-ideas/graffiti-advertising.html

http://www.wired.com/culture/lifestyle/news/2005/12/69741

http://www.medialifemagazine.com/news2005/aug05/Aug22/1_mon/news5monday.html

When to Use

- When targeting specific geographic areas down to neighborhoods.
- Generally to reach young people and street culture.
- Only if prepared to let artists create what they want while incorporating your required image.
- Great for consumer products, entertainment and event marketing.
- To get buy-in from community residents.

Graffiti - Permanent

Category: Offline

Frequency: Regularly

What is it

This graffiti is street level advertising that's used to bring a product or service into popular culture. Ideally, it's created by a real graffiti artist that incorporates your brand/message by spray painting directly onto a legally leased space in a location where graffiti is likely to exist.

Disney successfully used this advertising tactic to bring Mickey Mouse back into popular culture.

Pros

- Will draw attention if placed strategically and may be of interest to local media as a story in itself.
- Recruited graffiti artists will love it and tell their networks - their involvement will usually protect the images from being tagged over.
- Great photo ops for walkers by.
- May become known as a meeting place.
- Creates word-of-mouth.

Cons

- If it looks too much like advertising there could be backlash. Sony tried to use it to launch their PSP handheld in 2005. Other graffiti artists penned over many of the images. Problem - Sony told the artists what to draw.
- May require a permit.
- Space must be leased.

Photo: Captured at http://www.evergreensuk.com/nfl-pitch.htm

References & Examples

http://www.evergreensuk.com/nfl-pitch.htm

http://www.alwaysgreengrasspainting.com

When to Use

- Great for outdoor events or entrances in front of an indoor event.
- On a visible patch that the venue looks out over.

Grass Painting

Category: Offline

Frequency: Adhoc

What is it

Football is famous for painting logos on the field. Just watch any football game - particularly the "bowls" like the Superbowl and Capital One Bowl. Grass painting doesn't have to stay within the confines of a sports field. If you have an outdoor event or are looking for outdoor exposure, take advantage of the landscape. Paint your logo on a visible grasspatch like a hill or field within the range of audience exposure. It makes an incredible impact!

Most regular grass painters (yes, you can hire people to paint your grass green), can provide this service.

Pros

- Unusual, so really gets attention.
- If crowds are in viewer stands - you have a forced audience.
- Easily picked up by television and video cameras.

Cons

- Have to get permission from land owners.
- May have to rent the space.
- Can be susceptible to weather.

Photos: Captured from http://www.flashingblinkylights.com and http://sallygardens.typepad.com/photos/uncategorized/2007/08/05/herb_ice_cubes.jpg

References & Examples

http://sallygardens.typepad.com/photos/uncategorized/2007/08/05/herb_ice_cubes.jpg

http://spoutdirty.info/football-light-up-ice-cubes-multicolor-litecubes-brand-sku-no-10653.asp

http://www.epromos.com/product/8817297.html

When to Use

- Special hosting events and parties.
- To offer ice water to customers while they're browsing.

Ice Cubes

Category: Offline

Frequency: Adhoc

What is it

Want to generate talk at your next event? Place blinking ice cubes in the drinks. You can buy plastic ones in your event colors, or have them custom branded. These ice cubes come in a whole assortment of colors and designs like mini footballs and dice. Lots of fun! People love them.

Freezing brand colored flower petals into ice cubes is another way to flash your corporate brand - but make sure they're edible and not poisonous. They go over brilliantly!

Pros

- The unusualness attracts attention.
- Will create talk.
- Real ice cubes melt leaving the logo/item in the bottom of the glass or flowers floating on the top.
- The flashing cubes can be color branded and/or logo'd - people can take them home as momento.

Cons

- Short-term - the real ice cubes melt.
- Always have to be careful with flashing lights - can trigger seizures in a small percentage of people.

Photo: Taken in Richmond, BC by Charlene Brisson.

References & Examples

http://www.publiair.com/

http://www.inflatable2000.com/advertising.html

When to Use

- When holding a big sale or event.
- To draw attention to a location.
- When launching a new product.

Inflatables

Category: Offline

Frequency: Adhoc

What is it

You see great big recognizable blow-up inflatables on top of car dealerships (Michelin Man), Burger King (king's crown) and in front of McDonald's (coffee cup). There are thousands of different sizes and shapes for all size of businesses, large and small.

Many inflatables are generic and can be rented for the day, week or month.

It's critical that you keep the inflatable inflated - otherwise, you'll attract the wrong type of attention.

Pros

- Definitely draws attention.
- If placed up for a long period can become a landmark that people look for.
- Can be customized to pretty much anything you want.
- Can rent short-term.

Cons

- If set-up too long, becomes part of the scenery and loses its big-bang effect.
- Have to keep these inflated to the max.

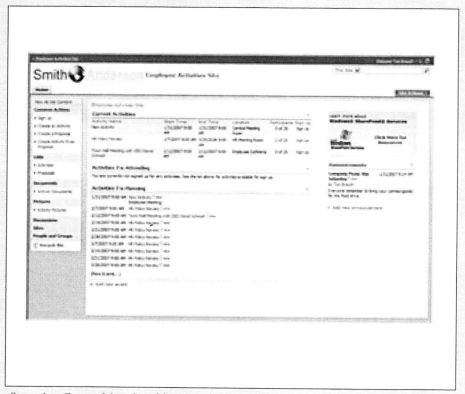

Screenshot: Captured from http://www.sharepointhostingprovider.com/sharepoint-demo/#campaign

References & Examples

http://www.engbsolutions.com/

http://sharepoint2010.microsoft.com

http://www.sharepointhostingprovider.com/sharepoint-demo/#campaign

When to Use

- With a business that has a significant employee base and/or a lot of remote contractors to keep in the loop.
- Great for businesses that have many locations.
- To facilitate and manage large projects with many contractors.

Intranet

Category: Online

Frequency: Regularly

What is it

Your internal customer - your staff and contractors - can be more important to market to than the outside world. If your staff isn't sold on your product or service, they'll never be able to convince others it's right for them.

An intranet is a website that is ONLY accessible to internal staff and contractors. It can be simple or very complex. Sharepoint is a good example of an intranet that facilitates communication and allows for the ability to market your products and services to your most important customer - your team.

Intranets are also being called business collaboration platforms.

Pros

- An intranet can be set up to promote AND communicate.
- Access can be customized.
- Files can easily be shared.
- Excellent for staffers and contractors that work off site.

Cons

- An intranet takes significant time and expertise to set up.
- Requires ongoing training and updating.

Photos: Winter Olympics Cowbell and Urban Spoon Apps taken by Charlene Brisson.

Pros
- Can enter your own establishment into many applications.
- Can encourage patrons to add reviews.
- People are more likely to try/buy from establishments reviewed by peers.
- Clever apps are downloaded by millions.
- Easy to find a capable programmer to build your own app.

Cons
- One can get overwhelmed by the number of applications and which to become part of.
- Can't control other's reviews.

When to Use
- To reach new customers.
- After you've established your core lead generators.

References & Examples
http://iphoneapplicationlist.com

iPhone / Smartphone Apps

Category: Mobile

Frequency: Regularly

What is it

There are thousands of ready-made iPhone and smartphone applications (also called Apps) that direct consumers to buy. Alternatively, to create brand awareness, you can always create your own clever application like Bell's Winter Olympic Cowbell.

Getting your business listed on many of the "find a location/business/restaurant/etc." applications backed up with positive reviews puts you in the path of available customers. Make sure that your business is listed on the most popular ones that make sense for your industry. In most cases, all you have to do is download the application and add your business. Yes, there are over 50,000 applications, so it will take some searching, listening, reading and asking to find out what your customers are into. Of course you'll also need an iPhone, iTouch or smartphone. Some of the interesting apps to get your business on are Find Green, Cellfire Mobile Coupons, mShopper, shopping.com, Urban Spoon, Find a Restaurant, Yelp, Clip and tens of thousands of others in every industry and genre you can think of.

Although iPhone is leading the way in applications, comparable smartphone apps are being released daily.

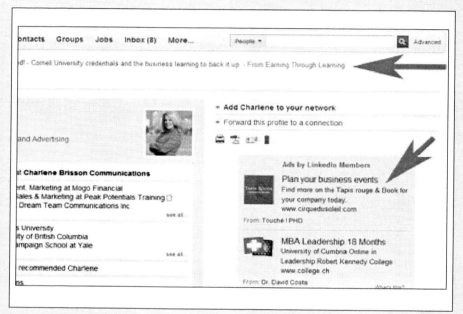

Screenshot: Captured at www.linkedin.com/charlenebrisson

References & Examples

http://advertising.linkedin.com/assets/Uploads/LinkedIn-SpecSheet.1209.pdf

When to Use

- To target a very specific business executive demographic and geographic audience.
- To drive traffic to your site.
- To increase brand exposure.

LinkedIn - Ads

Category: Online, Mobile
Frequency: Regularly

What is it

LinkedIn DirectAds work in a very similar way to Facebook ads. For as little as $10 per day you can run a 2 line 75 character ad with a 25 character headline and URL. Can create up to 10 variations. The ads are placed on LinkedIn profiles and throughout the LinkedIn network. Targeting is by geography, job function, seniority, industry, company size, gender and age. You can choose between paying only if someone clicks on your ad or for each one thousand impressions (CPM). For best results, make sure that your website or landing page deliver your ad promise.

LinkedIn also offers larger, more expensive ad campaign opportunities through custom polling, in-page video, expandable and display ads.

Pros

- Relatively inexpensive - can start with as little as $10 daily limit.
- Can zero in on your target market.
- Clickability makes it easy to drive traffic to your site/landing page.
- Easy to test and improve.

Cons

- Very easy to ignore.
- Sales conversion decreases if landing page doesn't speak to ad promise.
- Not all products/services have same conversion results.

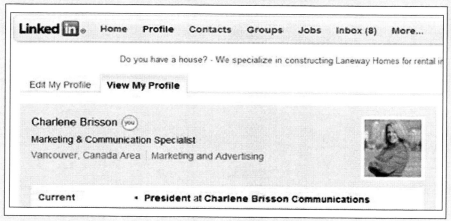

Screenshot: Captured at www.linkedin.com/charlenebrisson

Pros
- Basic profile is free.
- Considered a legitimate means of connecting with other professionals and to find business services.
- Assists with improving search engine rankings.

Cons
- Can get lost in the LinkedIn world without results.
- As with all social media, it may take a great deal of time interacting to get bites.

References & Examples
www.linked-in.com

http://www.linkedin.com/apps

https://www.linkedin.com/directads/start?trk=what&utm_source=li&utm_medium=ad&utm_campaign=what

When to Use
- When your product/service targets business executives or professionals.
- To build personal brand awareness and credibility.
- To connect with very targeted interest groups.

LinkedIn - Profile

Category: Online, Mobile
Frequency: Regularly

What is it

With 65 million members in over 200 countries and from all Fortune 500 companies (as of this writing), LinkedIn is the social media site used by professionals to network with other professionals. Often seen as a meca for job hunters, LinkedIn is also a terrific site to find other like-minded people who are looking for business solutions. Contractors do very well marketing their services here.

Free profiles are available. More advanced packages can be purchased to target and email LinkedIn members by geographic, industry and job type keywords. Recommendations/endorsements by other LinkedIn members provide viewers ome level of reputation comfort.

Thousands of discussion groups, one on pretty much any topic you are interested in, allow members to interact and promote business services. Can also start your own group to increase your profile on LinkedIn.

By going to "edit public profile" and then clicking on "customized buttons", you can download code that will let you put a click thru button to your LinkedIn profile on your website, facebook or blog.

Pros

- Easy to do segmentation marketing (ie - target a specific demographic). Here's some really unusual mag titles - Y'All: The Magazine of Southern People; I Love Cats; Cowboys & Indians; Garden & Gun; Bacon Business: The Pig Hunters Guide; Dance Teacher.
- Paid subscriber publications are more responsive than free and have a longer shelf life.
- Most publications will negotiate on rates - throw in free color, etc, in exchange for frequency.
- Inexpensive remnant space can be available last minute.

References & Examples

http://www.emagazines.com

Cons

- Frequency and repetition is important - one ad rarely does anything.
- Many magazine ad styles are full page "image ads" - direct response ads don't fit making ROI difficult to measure.
- Can be high cost with little return.

When to Use

- To capture a very targeted audience.
- For image advertising.

Magazines

Category: Offline
Online

Frequency: Regularly

What is it

Magazines come in all shapes and sizes. The distinguishing feature between newspapers and magazines is that most commonly magazines are published monthly and centre around a specific topic. They are usually available to purchase from newsstands or by monthly subscription. Some are distributed free door-to-door to targeted geographic areas.

- Ads come in all shapes and sizes and are measured by columns across and inches or lines high. Common sizes are double truck (two side-by-side full pages), full page, 3/4, half, 1/3, 1/4, 1/6 and business card size - all verticle or horizontal. You can purchase banners across the bottom or the top and across two or several pages depending on the rules of the magazine. Some magazines sell a verticle 1/3 page flap over the top of the front cover or a fold out off the front. Magazine ads are always more effective in full color and many don't even accept black and white ads.
- Advertorials look like editorial, but often differ in font and column size from the rest of the magazine. A disclaimer saying "This is an advertisement" always runs along the bottom or top so readers won't be tricked into thinking the ad is editorial.

Photo: Taken in Hong Kong by Carmen Choy.

Magazine Con't

Category: Offline
Online

Frequency: Regularly

Continued from Page 187

- A magazine supplement usually runs anywhere from 4 - 12 pages and is provided as final creative by the ad agency of the advertiser. Unlike newspaper supplements, the advertiser usually pays for the entire placement as an insert. The magazine may sell targeted advertisers around it. Supplements are useful if you want to do an overrun to hand out to customers at later dates. They are commonly used for openings, launches and to educate the public.
- Card stock pull-through inserts and smellys are also common in magazines. As are outbound inserts that require the publisher to plastic wrap the magazine so that your insert shows through to the recipient.

Many magazines are now offering online .pdf subscriptions to their publications. This offers additional opportunities to advertise on the website as well as in the .pdf pages of the magazine.

Pros

- Meetup.com has a built-in audience looking for events on your topic in your area - really easy to set-up.
- Inexpensive tools to set up offline and online meetings - usually a monthly charge.
- Online and offline meet-ups increase trust amongst your community leading to word-of-mouth.
- Announcing regular meet-ups gives you good reason to communicate with your list/customers.

Cons

- Often starts slow with few attendees - offline or online.
- Watch the limits on number of lines/attendees for online meet-ups. Unless things have changed, from my personal experience, gotowebinar.com CANNOT handle 1,000 lines and it can be disastrous for a live event - best to pre-record anyway.
- For the online tools, there is a definite learning curve.

References & Examples

http://www.meetup.com

http://www.gotomeeting.com

http://www.gotowebinar.com

http://www.superconferencepro.com/

When to Use

- As a lead generator.
- To build industry credibility.
- To build community around your product/service.

Meet-ups

Category: Offline / Online

Frequency: Regularly / Adhoc

What is it

Meet-ups are all the rage in segmentation marketing right now. These topic specific "get togethers" also help social media marketers transform anonymous connections into more meaningful face-to-face communities. Offline or online meet-ups are terrific lead generators. They provide your business with legitimate opportunities to bring like-minded customers and/or prospects together while you, as the credible expert, share information and teachings in order to build a following and eventually sell product/services. Keep events soft-sell or no-sell for best results. The goal is to get people to know you and trust you as a leader in your field. The actual sales will come later. Having your books on sale and brochures out is a good strategy.

Dedicated online platforms make it really easy to put groups together. www.meetup.com coined the phrase and claims to have 6.1 million members and 2,000 groups meeting daily. The system is easy and inexpensive. Users often charge attendees a minimal fee to cover rental of the room and basic refreshments for the meet-up.

Other online technology like www.gotomeeting.com; www.gotowebinar.com; and www.superconferencepro.com provide relatively simple tools for entrepreneurs to gather groups online for regular and adhoc topic-specific meetings.

Photo: Captured at http://www.gomobileadvehicles.com

References & Examples

http://www.altterrain.com/AdverLive_live_billboard_advertising_mobile_truck.htm

http://www.outdoorvisibility.com

http://bulldogbillboards.com

http://www.gomobileadvehicles.com/index.php

When to Use

- To launch a product.
- In place of, or as part of a transit or outdoor campaign on targeted routes.
- As part of product sampling station.
- As a component of "spot the mobile billboard" and win contest.

Mobile Billboards

Category: Offline

Frequency: Regularly Adhoc

What is it

WOW! These are trucks driving around with dynamic advertising (usually video) on the sides and/or the back. Real eye-catchers at night as they make their way through town and down targeted routes in illuminated splendor.

These mobile billboards are also very effective when set-up near sampling stations with street teams.

Pros

- Unusual so very eye-catching.
- Can be used in many ways.
- Easy to include an interactive contest component.

Cons

- Difficult to measure ROI if no interactive component.
- Call-in and texting interactivity a challenge now that it's illegal to call/text and drive in most locations.
- Not available in all city centres.

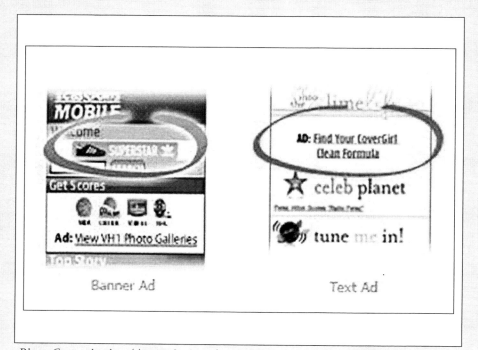

Photo: Captured at http://www.admob.com/

References & Examples

http://www.admob.com/

http://www.mojiva.com/?ap=69372

http://about.blyk.com/

When to Use

- To build opt-in list.
- To launch new product.
- Expand awareness.

Mobile Marketing

Category: Mobile

Frequency: Regularly

What is it

The newest form of advertising in Mobile Marketing is called contextual mobile phone advertising. Companies that have robust networks of mobile website publishers provide you the opportunity to run text and banner ads on their networks' mobile websites that match up with your keywords - very similar to Google Adwords. You pay per click based on keyword bidding, although some entertain charging by thousand impressions (CPM) as payment for larger buys and organizations.

Some telecommunication companies offer a specific target audience that your ads are sent to and if the audience finds your offer appealing, they click thru. In return for their click thrus the audience will get free texting and calling discounts, similar to Paid-to-Read tactics.

Pros

- Can target your audience using keywords.
- Because it's a new medium, less advertisers are using it, so less competition in the market.
- Only pay if someone clicks on your ad.

Cons

- Not clear on effectiveness.
- How serious are the people who click thru using their cell phones?

Photo: Captured at http://www.altterrain.com/Mobile_Video_Billboard_truck_advertising.htm

References & Examples

http://www.altterrain.com/
Mobile_Video_Billboard_
truck_advertising.htm

When to Use

- To launch a product.
- For image advertising.
- To sell entertainment, theatre, venues, etc.

Mobile Video Cubes

Category: Offline

Frequency: Regularly
Adhoc

What is it

These illuminated cubes sit on the roof top of a car, van or SUV and run dynamic video. As the vehicle (usually a taxi cab or consumer transportation vehicle of some sort) travels around, the ad/s keep running one after the other or again and again. Each side of the cube (all four) features an individual screen.

Pros

- Really unusual to see in most centres so they attract attention.
- The cube is large enough for people to make out what is on the screen.
- Can replace or complement transit ads.
- They look great at night.

Cons

- Only available in select large cities.
- Difficult to focus on a specific route as taxis usually travel far and wide.

Photo: Ad appearing in small town theatre taken by Charlene Brisson.

References & Examples

http://media.cineplex.com/OnScreen/DigitalPreShow.aspx

When to Use

- To launch a new product.
- For image advertising
- To reach a very demographically and geographically targeted audience.
- To associate yourself with a particular movie and/or star.

Movie Ads

Category: Offline

Frequency: Regularly Adhoc

What is it

As a revenue stream, movie theatres, both chains and independents, run 30 and 60 second spots before the trailers. Some are customized for the theatre audience, others not so much. The big chains historically run big budget national ads. Independent theatres will take on local ads for both businesses and not-for-profits. Many local theatres loop static ads (see left). So the advertiser need only provide a jpeg of the ad - very simple. On average 3 or 4 ads run prior to the trailers.

Pros

- Provides one of THE most captive audiences available.
- Movie goers have warmed up to this type of advertising.
- Can do micro-targeting.

Cons

- Many movie goers are hostile toward advertising - but they're getting used to it.
- If the movie tanks, the audience won't be there, hence limited impressions.
- Can be very expensive to hook-up with a block buster film.

Photo: Captured at http://www.rhinoink.ca/murals/muralads/koodo.html

References & Examples

http://www.rhinoink.ca/murals/muralads/koodo.html

http://www.altterrain.com/graffiti_advertising.htm

When to Use

- To create word-of-mouth.
- To make a large impact.
- For image advertising.
- To target geographically.
- To cover up a traffic facing unpleasant looking exterior wall.

Murals

Category: Offline

Frequency: Regularly
Adhoc

What is it

A mural is art that is painted directly onto an interior or a leased exterior wall, although we mostly see them on the outside of buildings. An advertising mural combines art with an advertising message that aims to catch the attention of pedestrians, passing vehicles and transit traffic.

Murals may be hand or spray painted. Although similar, they are not quite as "wild" as graffiti art and the creative is totally controlled by the advertiser.

Pros

- Can be very attention-grabbing with the right creative.
- Can be seen as an art piece, and depending on the creative, can improve the neighborhood.

Cons

- If up too long, the mural eventually becomes invisible to passers-by.
- Will have to lease the building wall space.
- May be target of graffiti.

Photo: Taken in Vancouver, BC at F5 by Charlene Brisson.

References & Examples

http://videonametag.com/

http://www.mbd2.com/badge.htm

http://www.digitalnametag.com/

When to Use

- At trade shows and events to draw attention.
- As a networking tool.

Name Tags - LED

Category: Offline

Frequency: Regularly

What is it

I LOVE these! Digital name tags with an electronic sign scrolling your name in lights across your chest on a 4`x 3`card. They're fabulous as they can't be missed no matter how hard one tries!

Just before going to print I captured a fully digital LED name tag (at left) at a conference which was rotating high resolution photos and graphics. It's called a video name tag and comes with a wrist holder so you can wear it on your chest or your wrist. WOW!

Pros

- Really attracts attention.
- You'll probably be the only person/company wearing them.
- Excellent conversation starter.
- Can program the text/graphics yourself.
- Very inexpensive for the impact you'll get.

Cons

- Much more expensive than regular name tags.
- Have to change the batteries fairly often.
- Shy staff will complain - they shouldn't be your trade show/event staff then.

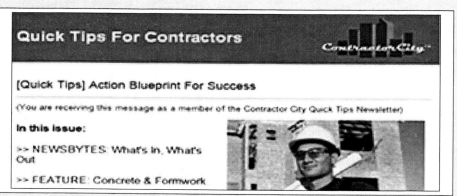

Screeenshots: Captured from http://www.contractorcity.com newsletter email.

Pros
- Inexpensive to create and send.
- May reach thousands or even millions of online newsletter subscribers.
- Limited advertising space exposes readers to fewer competing ads.
- Less pressure to print out newsletters, can go fully online with a pdf downloadable version or online newsletter service provider.
- Easy to set up electronic newsletters through service providers.
- Can archive on your website for extra value content.
- Opportunity to include customers and suppliers into the content.

Cons
- Easy to ignore or delete.
- If too blatantly salesy, people will not read it.

When to Use
- To target by interest.
- To keep customers and prospects informed and engaged.
- To educate.

References & Examples
http://insidestoreadvertising.com/newsletter_ads

http://www.contractorcity.com

http://www.getresponse.com

http://www.constantcontact.com

Newsletter

Category: Offline / Online

Frequency: Regularly / Adhoc

What is it

Many corporate newsletters are in themselves advertising pieces that provide product/service information, related issue education, user reviews, recommendations, updates, product launches and any other interest related topic. All is in the effort to 1) remind the current customer that they have made the right purchasing decision; 2) upsell the reader to other products/services; and 3) to sell prospects. Advertising is often placed within newsletters and they offer a terrific opportunity to include joint venture partners as a return for your content in their newsletter.

Although some organizations still print out copies of their newsletters, most have migrated to electronic versions. Similar service providers used for email blasts like www.GetResponse.com or www.ConstantContact.com make it easy by providing electronic drag and drop templates and simple list management and readership tracking.

Pros

- Can make a bigger impact with smaller ads as usually only 25-30 pages and page size is typically smaller than a daily tab.
- In most major cities.
- Some of the commuter papers are audited which means that their distribution/readership numbers are confirmed by a third-party auditor.
- Can negotiate reduced rates when advertising in multiple markets.
- Some commuters are owned by daily papers or other media companies so can negotiate a reduced rate combo media buy.
- Most commuters have online editions so you can also get online exposure.

Cons

- Watch for unaudited commuter papers as they cannot 100% guarantee their distribution/readership.
- Not all cities have commuter papers, for example, Metro is only in 3 US cities - New York, Boston and Philadelphia.
- Commuters are getting very competitive in the big cities - lots to choose from, so the rates can be very reasonable - just watch for the distribution channels and how they guarantee their distribution/readership.

References & Examples

http://www.metro.lu/about

http://tonightnewspaper.com/

http://tinyurl.com/148Ways-Newspaper-Commuter

When to Use

- In combination with an integrated campaign.
- When you have a limited print budget.
- To go after a 25-45 demographic.

Newspaper - Commuter

Category: Offline / Online

Frequency: Regularly / Frequently

What is it

Commuter papers are distributed free of charge by carriers on street corners; on transit such as some subways, express trains; in newspaper boxes; strewn around coffee shops; in subway and skytrain stations. The Metro commuter paper is published in many major cities across the US and Canada. Its format is smaller than a typical daily tab newspaper and usually only 20 to 30 pages. Generally, the commuters target 25-54 year olds on their way to work with quick snippets of news and entertainment. In many locations, afternoon commuters are now being produced.

Metro claims to be the largest global commuter newspaper, publishing in 100 cities around the world. Their target is "millions of industrious metropolitans [who] take to the streets, buses and trains on their way to work. They are young, well networked trend-setters, cash-rich but time-poor, with healthy media appetites and perpetually shifting tastes" (http://www.metro.lu/about). This is a typical demographic target of commuter papers.

See "Newspapers - Daily" on page 211 for the a list of ad sizes and types that can be purchased in commuter papers.

Photo: Taken by Charlene Brisson.

References & Examples

http://www.communitymedia.ca/industry/

http://classified.van.net/classified/classified.nsf/index

When to Use

- To keep in front of your community audience.
- As a component of an integrated campaign.
- To secure sponsorship for an event.
- To gain credibility within the community.

Newspaper - Community

Category: Offline / Online
Frequency: Regularly / Adhoc

What is it

Community newspapers are most often delivered free of charge to households on a weekly or bi-weekly basis. They are primarily filled with ads from local retailers and community not-for-profits. Editorial is locally focused on people, events and politics. It is how families mostly stay in touch with what's happening in their neighborhood. Often these papers are jammed full of inserts and flyers. Community papers offer the same advertising opportunities as daily papers, but at a much reduced cost. See "Newspapers - Daily" on page 211 for the a list of ad sizes and types that can be purchased in community papers.

Pros

- Can reach a very specific audience.
- Can be "the place to go" for what's happening and what's on sale.
- Most are online as well and can purchase bonus online exposure.
- Relatively inexpensive.

Cons

- A lot of free community papers end up in the recycling box.
- Newspapers have been losing readership for the last two decades.

Pros

- Can get overruns of supplements.
- Because of declining readership, newspapers are more likely to negotiate a deal - almost always for color and for long term ad buys. Ask for bonus promotional space.
- Can get discounts when advertising in more than one paper in a chain.
- Last minute remnant space is often available at a significantly reduced rate. This is when last minute advertisers pull out for a number of reasons.
- Paid dailies still have a fairly solid older demographic.

Cons

- Some newspapers insist you buy extra copies for promotional purposes so you'll give them away and they can pump up their auditing numbers.
- For the past decade newspapers have been losing readership.
- Impossible to track exactly how many people have actually seen your ad.

References & Examples

http://www.nationwideadvertising.com

http://www.usnewspapers.com/

http://www.cna-acj.ca/en/aboutnewspapers/media/canadian-dailies

http://www.pressdisplay.com/

When to Use

- When you want frequency.
- When you have a large budget.
- For image advertising.
- As part of an integrated campaign.

Newspaper - Daily

Category: Offline / Online
Frequency: Regularly

What is it

Newspapers come in two basic sizes: broadsheet and tab and are moving to the internet in record numbers for subscribers to view in pdf format (http://www.pressdisplay.com). All major cities have at least one daily (often two) plus one or two weekly community papers. The daily papers usually come in two sizes:

- **Broadsheet size** newspapers historically target a white collar more mature 35 - 65 and sophisticated reader. Examples are the New York Times, LA Times, Washington Post, Toronto Star, Vancouver Sun, etc.
- **Tab size** newspapers often target a younger more blue collar reader 25-45 and commonly have a large sports section.

Here are some newspaper advertising basics:

- **Display ads** come in all shapes and sizes. They are measured by columns across and inches or lines high. Common sizes are double truck (two side-by-side full pages and/or banners), full page, 3/4, half, 1/3, 1/4, 1/6 and business card size - all verticle or horizontal. You can purchase banners across the bottom or the top, depending on the rules of the paper. Black and white, spot color and full color are options. Always try to add at least one spot color - red is best to get a reader's attention. Right hand page, right hand bottom corner is the best position on a hard copy, even if the rep tells you it isn't - I haven't

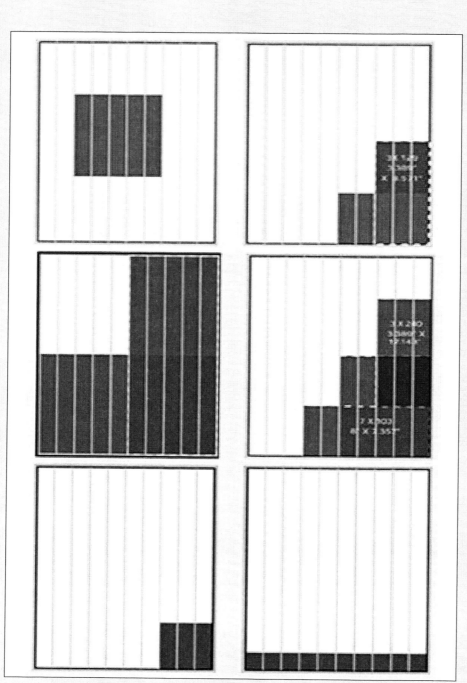

Screeenshots: Captured from http://www.png.canwest.com/sunspec.html

Newspaper - Daily con't

Category: Offline / Online
Frequency: Regularly

Continued from page 211

seen any research yet on electronic ad positioning.

- **Flex form ads** are the most recent iteration of the standard print ad. Instead of being another type of square, the editorial actually wraps around the unusual shape of the ad - and sometimes the ad will interject into the editorial. Readers can't help but notice the intrusion into editorial or the unusual shape.
- **Advertorials** look like editorial, but are often in a different font and column size than the the rest of the newspaper. A disclaimer saying "This is an advertisement" always runs along the bottom or top so readers are not tricked into thinking it is an unbiased article.
- **Supplements** usually runs anywhere from 4 -16 pages. It is filled with editorial often provided by the major advertiser/ sponsor - or if the supplement is large enough, the newspaper will provide a writer. Newspapers often require a minimum of 50% advertising and you as the primary advertiser/sponsor must either sell the ads yourself or provide a list of people who will purchase ads. Supplements are useful if you want to do an overrun to hand out to customers at later dates. They are commonly used for openings, launches and educating the public.

Photo: Captured from http://ri.rediffiland.com/homepimages/home8/756/88f35efdfa5904e1c4a8e4e60463 04cc/homep/images/1252423782

References & Examples

http://getpaid-ptr.info/

http://www.bugonptc.com/paid-to-read-advertisements.pdf

When to Use

- If the PTR company can guarantee a targeted list.
- To test other potential ad methods when you've run out of successful ways to secure customers.
- To reach an active smartphone user target.

Paid-to-Read (PTR)

Category: Online, Mobile

Frequency: Adhoc

What is it

PTR, also called Paid-to-Click advertising is offered by online companies that register affiliates or members. These affiliates are presented with a number of ads on a regular basis through email and/or text and are paid every time they click on an ad that is of interest and spend the required amount of time on the subsequent website. Of course the affiliates are paid less than the advertiser pays so that the online company can sustain the business. There have been many scams around online PTR, so some caution is advised.

There are mobile operators who offer a similar service but rather than paying cash, they pay affiliates free text messages and minutes in return for clicking on mobile messages.

Pros
- Relatively inexpensive.
- You only pay for the clicks/reads.

Cons
- Affiliates are generally clicking for pay rather for true interest.
- Affiliates are often "sign-up" people, so your opt-in list may get clogged with non-buyers.
- Be cautious of unprofessional PTR companies.

Photo: Captured from http://www.parkingstripe.com/index.htm

References & Examples

http://www.parkingstripe.com/index.htm

http://www.aapglobal.com/aapstripes.php

When to Use

- When your product is available in a store that features stripe advertising in their parking lot.
- For image advertising.

Parking Stripes

Category: Offline

Frequency: Regularly

What is it

Even the stripes on privately owned parking lots are available to advertise your message on. If your target market shops in a specific mall, why not advertise on the parking stripes so your product/service will be top of mind while shopping.

While writing this book, parking stripes with sound have been launched (http://www.parkingstripe.com/news.php?itemid=27). Run your jingle or call to action. This can be an incredible opportunity for advertisers, although it's yet to be determined how the public will respond.

Pros
- Captured target market in parking lots.
- Easy to target your customer.
- Very unusual so will attract attention - the first time it's seen anyway.
- Can make an offer to track ROI (use this code at the cash register and get 20% off, etc.).

Cons
- Do people really see the stripes when they get out of their car?
- Limited space to put your message.
- Can be difficult to measure ROI without an interactive component.

Pros

- Low cost per impression - can get in the game with very little budget.
- Easy to test ad copy and keywords for best results.
- Can target micro-niches at a lower cost producing higher conversion results.
- One of the best online means to drive qualified traffic to your site.
- Can set up tracking to follow results.

Cons

- Must adhere to limited ad size - 25 bolded title characters and 35 description, plus URL.
- Searchers prefer to click organic, not paid listings.
- Side listings easy to miss.
- Paid rankings listed on page two or more are rarely seen.
- Some keywords are very competitive, so it's difficult to get a paid ranking at a low budget.
- Know your keywords - driving the wrong traffic is costly.
- Campaign set up, bidding and tracking can be complex.

References & Examples

http://tinyurl.com/yfwz5fl

https://adcenter.microsoft.com/

http://advertising.yahoo.com/?o=USPX06

When to Use

- To establish online brand credibility.
- To increase website traffic.
- In combination with other internet advertising.

Pay-per-Click (PPC)

Category: Online, Mobile

Frequency: Regularly

What is it

Pay-per-Click (PPC) is the advertising platform that search engines use as their major revenue stream. It is named as such because you as an advertiser only pay if people click through on your ad which links directly to your website or landing page. If nobody clicks, you don't pay. Plus, you can set maximum budgets per day for your campaigns so you will never go over your budget. Of course, the more you are willing to pay-per-click, the higher up in the paid listings (called rankings) your ad will appear.

It's all based on bidding for popular keywords (popular for your product or service that is). You set a maximum bid per click to secure your search engine ranking positioning. What this means is that your 3 line text ad (including the headline) plus photo and URL will show up either on top of the list above all unpaid organic rankings or down the right hand side of the page.

These Pay-per-Click ads are identified to the user by being just a little bit different than the organic listings. Google and Yahoo identifies the paid search listings at the top of the page by having a lightly screened yellow background and the words "Sponsored Link/s" or "Sponsored Results"

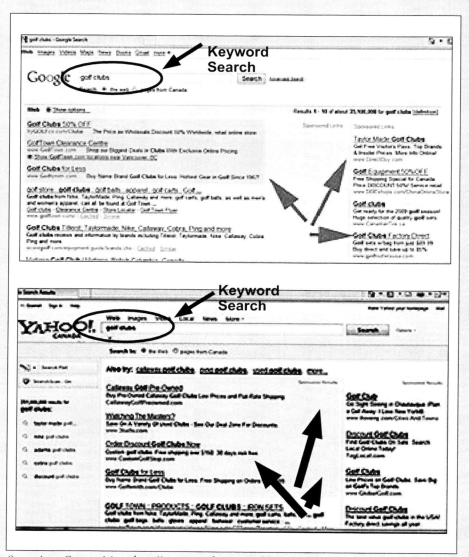

Screenshots: Captured from http://www.google.com and http://www.yahoo.com

Pay-per-Click (PPC) - Con't

Category: Online, Mobile

Frequency: Regularly

Continued from page 219

to the right of the top listings and as the header for the paid listings along the side of the page. Bing has a light blue color screen behind the top paid listings and the words "Sponsored Sites". To see the colored background, you may have to tilt your monitor - it can be difficult to see immediately, which is strategic by the search engine to try to blend the paid and unpaid organic listings. This is good for you the advertiser, as research shows that searchers are less likely to click on the paid ads.

PPC campaigns are self-service. You can set up campaigns yourself, enter your keywords and select where you want your ads to appear such as in the search listings, on the search engine blog sites, websites, email (gmail) and/or YouTube (in the US). The more complex you get with the campaigns, the more likely you're going to want to hire someone experienced to manage and optimize your PPC campaigns.

As with all landing pages, make sure that you deliver what you promise in your PPC ads. In most cases, landing pages outside of your home page are critical to align targeted messaging with ad groups.

Photo: Captured from http://www.aapglobal.com/aapmotion.php

References & Examples

http://www.aapglobal.com/aapmotion.php

When to Use

- In high foot traffic areas - excellent in airports.
- For image advertising.

Pedestrian Motion Panels

Category: Offline

Frequency: Regularly

What is it

Very clever! This signage is only in motion when the person passes by. In fact, nothing is moving - yet as the viewer passes, the individual panels give the appearance of being animated. Great for large high foot traffic areas like airports, convention centres and underground subway connector tunnels.

Pros

- Very unusual so catches attention.
- Very large in size, so difficult to ignore.

Cons

- Viewers must be looking at the sign while they are moving to get the full effect.
- Can be costly.

Photo: Captured from http://www.adaplacemat.com/component/jce/?tmpl=component&task=popup&img=images/stories/samples/jsm_camb_med.jpg&title=jsm_camb_med&w=1600&h=1035

References & Examples

http://www.adaplacemat.com

http://www.theplacematpeople.com/

When to Use

- To promote your own products and/or events.
- For product launch.
- As a suggestive selling tool.
- To get in front of a targeted audience.

Placemats

Category: Offline

Frequency: Regularly
Adhoc

What is it

Placemat advertising is quite common in restaurants and food fairs. Some businesses, like McDonald's use their own paper placemats to promote their latest product of the month, toy and/or characters.

To pay for the placemats, some local eateries use mats that display advertising from independent local businesses.

Pros

- Very inexpensive.
- Relatively captive audience.
- Can be seen as part of the community.
- People read placemats when they're bored or dining alone.

Cons

- Easy to ignore.
- The placemat gets covered by dishes and food.
- Questionable ROI.

Pros

- Can introduce an audience to your products through education - they'll be more open to receive your message.
- Easy to create.
- Very inexpensive.
- Keeps people coming back to your website.
- Using syndicates can provide credibility.
- Can insert sales messaging at front and back of podcast.
- According to Edison Research, about 80% of podcast subscribers are most likely to take action on a podcast advertising or sponsorship if they like the host.

References & Examples

http://podcast.com/

http://www.apple.com/itunes/podcasts/specs.html

http://tinyurl.com/148Ways-Podcasts

Cons

- Lots of podcast competition.
- Each recording has to be interesting or people won't listen to any others.
- First time iTunes upload can be challenging.

When to Use

- To keep people coming back to your website.
- To build brand credibility.
- To educate your customers and prospects.
- To target by interest.
- To soft-sell your products/services.
- As content delivery tool for membership sites, weekly tips, etc.
- As Q & A session recordings.

Podcasts

Category: Online, Mobile

Frequency: Regularly

What is it

Podcasts are digital audio files that contain information content for your customers and prospects. Often educational as topic-specific information presentations or recordings of interviews, the content can be any audio that is of interest to your audience. Posted on your website, usually in an mp3 file format, a podcast can be easily downloaded by users onto their computer, an ipod or mp3 player for future listening. Podcasts can also be listened to right off of your website. There are syndicates like iTunes, that will accept your podcast for distribution free of charge or for a small fee.

Creating a podcast is very easy using a microphone and recording directly onto most computers. Once your podcasts become popular, you can insert mini ads up front and at the end along with recommending products and/or services.

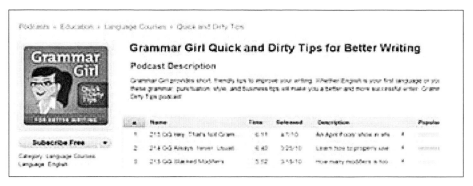

Screenshot: Grammar Girl free podcasts captured from iTunes.

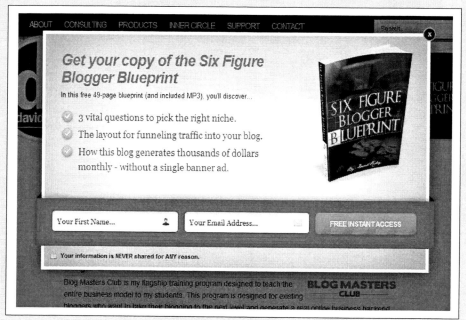

Screenshot: Captured from http://www.davidrisley.com

References & Examples

http://www.pep.ph/advertisement.php

When to Use

- To catch attention of viewers.
- To add a dynamic element to an existing banner ad.
- For list building - using pop-up or info capture.

Pop-up Ads

Category: Online

Frequency: Adhoc

What is it

Pop-ups are ads that launch/pop-up in either a small window or a new tab on your internet browser when entering a website; viewing other pages on a website; or rolling over an online display ad. Pop-ups are an uncertain way to advertise since most browsers are pre-set with pop-up blockers.

Pop-ups are often used for list capture. A small "sign-up now" window will pop-up, scroll-up, scroll down, or magically appear on the screen. Software is available for you to set a list data capture pop-up on any website and/or page. I'm not willing at this point to share links on this software as I haven't found a company that doesn't make you jump through hoops or become a lifetime member to get the info.

Pros

- Can divert web visitors to your website.
- Provides interactivity on your site.
- Is an attention getting list building tactic.

Cons

- Easy to ignore.
- Are seen as irritating and often blocked by pop-up blockers.
- Have been used in seemingly unethical ways by questionable advertisers.
- May interfere with viewers' online activities.

Photo: Captured from http://adage.com/article?article_id=138704

References & Examples

http://adage.com/article?article_id=138704

http://trendwatching.com/trends/POPUP_RETAIL.htm

http://tinyurl.com/148Ways-PopUpStore

http://tinyurl.com/148Ways-PopUpRestaurant

When to Use

- When entering a new market.
- Launching a new product.
- To achieve high impact brand visibility.
- In high-traffic pedestrian locations.
- To take advantage of high sales season.

Pop-up Store

Category: Offline

Frequency: Adhoc
Seasonally

What is it

This is a lot of fun for pedestrians. A temporary store just pops-up one day in a high-traffic location with excessive signage, enticement and activities. The pop-up store is usually not around for more than 3 months. The store generally has heavy signage and excessively branded and enthusiastic staff. Recently the trend has caught fire with restauranteurs - they're popping up all over the world.

Excellent for the Christmas period when shopping (and eating) is at its peak.

Pros

- Makes a major impact.
- Short-term commitment.
- Lots of available short-term retail space in high-traffic areas during recessionary times.
- Landlords are willing to negotiate price and term while finding long-term tenant.
- Low cost marketing.

Cons

- The store itself may not be financially profitable over short-term.
- Space may be empty because of low-traffic or other relevant reasons - research, research, research.

Pros

- Great relationship building if personalized. The customer notices and appreciates the extra effort and feels you "care."
- Quick and easy.
- Very inexpensive to produce and cheaper to mail than an envelope.
- Can easily integrate personalization through direct mail houses.
- Can be designed outside of the standard size.
- Can do everything online - no need to spend time with a custom printer.
- With online printers can print small quantities, use templates, add themes and logos.
- Good branding opportunity.

References & Examples

http://www.mypostcardprinting.com/

http://www.vistaprint.ca

https://www.sendoutcards.com

Cons

- Unless the postcard is specialized and/or personalized in some way - it gets tossed.
- Can look unprofessional if not carefully designed.
- Check with the Post Office before designing to make sure that you are within the self-mailer standards - if the card is too small, you'll have to put it in an envelope - if it's too large - you'll have to pay more to mail it.

When to Use

- As a reminder of an upcoming important date such as a payment, special sale or a product a customer has been waiting to arrive.
- As a thank-you for business.
- To build into your CRM communications.

Postcards

Category: Offline

Frequency: Regularly

What is it

Traditionally, many of us think of the postcards we send to our friends while vacationing. A standard rectangular size with a glossy image on one side and place to write a short message (wish you were here) on the back along with the name and address of the recipient. No envelope needed - just a stamp. Postcards can be used as effective tools for business owners to communicate to their clients with personalized or automated messaging. With the move to everything digital, a personalized, or seemingly personalized postcard can get the attention of your customer or prospect unlike anything else. Standard size is 4" x 6", but go up to 7"x 10" and everything between.

Some online companies will print and mail your campaign. Other companies like Send Out Cards will capture your handwriting and card image. You enter your message online, they print it onto your customized card, stamp it and mail it. A colleague of mine received one of these personalized cards from a sales person he spoke to two days before. He was so shocked to get a hand written card that he passed it all over the office to be examined and see if it was actually hand written - we couldn't tell. Brilliant.

Postcards can also be used as mini brochure handouts at your location or as rack cards, so design accordingly.

Photo: Captured from http://shop.majisign.co.uk/acatalog/posters_flyers.jpg

References & Examples

http://www.lynchposters.com/HTML/Sizes.htm

http://shop.majisign.co.uk

When to Use

- To promote an event or a one-off sales event.
- To make a quick display window change or update.

Posters

Category: Offline

Frequency: Adhoc

What is it

Posters have been used for promotion since the invention of paper and writing. Any paper or cardstock containing an image and/or text used to promote an event, one-off sale, service or product can be called a poster. You see them in store windows; on the walls of public buildings; on school/work/community centre bulletin boards; telephone poles and even around cash registers. The size is usually from 8.5"x 11" to 27"x 41" - but can be as large as your printer can print without becoming a billboard.

Pros

- Can provide a quick means of getting your message out.
- Inexpensive way to change your window messaging.
- Simple and bold messaging can attract walk and drive-by attention - i.e. "SALE 50% OFF."
- For non-sales events, allows for a combination of graphics and enough text to result in decision making/action taking.

Cons

- If creative is too busy, won't be seen.
- Have to be replaced regularly due to 1) theft; 2) weather damage; and/or 3) being vandalized.
- To get posters up outside of your location, in most cases you will require approval from building or location owners/operators.

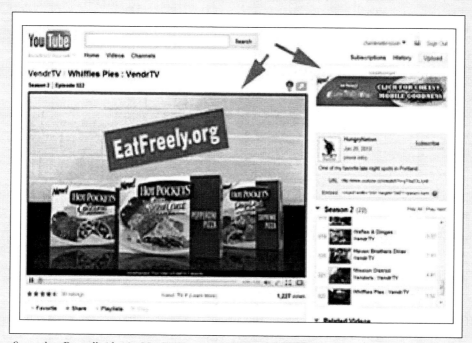

Screenshot: Pre-roll video for Hot Pockets from www.YouTube.com.

References & Examples

http://adage.com/digitalnext/article?article_id=142145

http://tinyurl.com/148Ways-Pre-Roll

When to Use

- As a component of a big budget integrated campaign when using large networks.
- When your product relates to the news story/video.

Pre-roll Video

Category: Online

Frequency: Adhoc

What is it

Pre-roll video is pretty much a short traditional television ad that the viewer is forced to watch just prior to a video, news story, video game or other website service. It is often used by large online venders such as MSNBC and select YouTube videos, but is becoming the norm. The ad is a give-and-take for online viewers. The viewer can freely watch the content, all they have to do is make it through the 5, 15 or 30 second ad first. Stats show that 1 in 6 people will abandon watching the video if they have to watch an ad first. Hulu.com is showing promise by giving viewers a choice of ads to watch.

On YouTube, advertisers can combine a display ad for the same product with the pre-roll - to the right of the video window - that doubles impact to the viewer.

Pros

- Forced viewership in exchange for rich content.
- The more popular the content, the more impressions.
- Can be keyword targeted.

Cons

- Some viewers see the ad and immediately abandon.

Screenshot: L'Oreal contest on 99.9 Radio at http://toronto.virginradio.ca/contest/97307

References & Examples

http://toronto.virginradio.ca/contest/97307

http://ellen.warnerbros.com/giveaways/

When to Use

- To enhance an ad buy.
- To make an impact with your target market.
- To generate word-of-mouth.

Prizing - Media

Category: Offline / Online
Frequency: Adhoc

What is it

The media (radio, TV, newspapers, magazines and media websites) are always looking for great prizing and give excellent exposure in return. On-air and in-paper giveaways help to keep their listeners and viewers tuned in and reading, so if you have a product or service that relates to their audience, offer it up. The bigger the quantity, the higher the value, the better. With the larger media, most only accept prizing from existing advertisers, but smaller markets may make exceptions if the prizing is fabulous.

Rosie O'Donnell started a trend with day-time talk shows of giving everyone in the audience a gift - now most are doing it which opened up a huge opportunity to provide prizing.

Pros

- People are more interested in giveaways than ads.
- Great prizing gets frequency and is usually given away during prime time hours and in best read sections.
- You don't have to be a giant firm to provide a prize - just an advertiser.

Cons

- May have to run a paid ad campaign to get your prizing accepted.
- Can be costly to give away a lot of product or very expensive product - one or two small items won't be wanted.

Photo: Captured at http://www.4imprint.ca/product/C101039/Death-by-Chocolate-Gift-Basket

References & Examples

http://tinyurl.com/25qhloh

http://www.vancouverfringe.com/gala-auction-items/

http://www.realauction.ca/

http://pages.ebay.ca/charity/

http://givingworks.ebay.com/

When to Use

- To build relationship with organizations whose membership match your target audience.
- To show goodwill in the community.

Prizing - Not for Profits

Category: Offline / Online
Frequency: Adhoc

What is it

Many not-for-profits, local sports teams, church groups and elementary schools hold fundraising raffles and galas and are on the lookout for prizing. You can provide your product or service as a prize in exchange for exposure (also called in-kind), which can be a logo or listing on the tickets; in the program; in advertising; a logo and link on the organizers website; and of course you'll get the attendees vieing for your item/s at the event's silent auction.

Many organizations are also running online auctions to raise funds - another great opportunity to get exposure. In fact eBay and others have easy auction set-up tools for charities to use. Look there for the right fit to give away your products.

Pros

- Great exposure for your product/service.
- Less expensive to provide product than cash.
- Seen as good corporate citizenship.
- There are many events/auctions to choose from.

Cons

- Make sure the raffle (outside of an event) is registered - otherwise in some areas it may be illegal.
- Giving once can often start a flood of requests to support other groups.

Photo: The Insider television show using an iPad taken by Charlene Brisson.

Pros

- Creates top-of-mind awareness among customers/clients.
- Creates word-of-mouth and water cooler conversation.
- Can get an instant following based on the association with the star or show.
- Many independent films also use product placements which are much less expensive than the big studios.

Cons

- Can be risky, as you never really know how popular the project will be - unless a guaranteed block buster sequel.
- Very expensive for block buster films/programs.
- Too obvious product placements can be a turn off.
- Too subtle product placements may go unnoticed by viewers.

References & Examples

http://tinyurl.com/148Ways-Product-Placement

http://www.filmbrandconnections.com/home

http://tinyurl.com/148Ways-Gaga-Telephone

When to Use

- To boost brand awareness.
- To associate with high profile/well known stars.
- To reach a very specific audience.

Product Placement

Category: Offline

Frequency: Adhoc

What is it

Product placements are becoming the norm in movie and television series culture. Advertisers pay to have products and/or logos subtly placed within a scene and/or are talked about by the characters as part of their lifestyle.

A not so subtle product placement example is in the movie "I am Sam." The lead character played by Sean Penn works in a Starbucks. The movie opens with him doing his job in a Starbucks location wearing a branded apron and stacking branded cups. Another is the in the "Tooth Fairy" movie. Many of the scenes are in a hockey arena and just like in real life, the rink boards feature advertising. Coke is the most prominent. The Castaway is another blatant product placement example - Tom Hanks character works for Fed Ex and the Fed Ex cargo plane crashes - logos are everywhere.

Much more subtle placements include everything from computers, vehicles, drinks (alcoholic and non-alcoholic), designers and chocolate bars to Jimmy Choo's (shoes) in Sex in the City. There really is no limit.

Apple's new iPad is showing up as props in a whole range of TV shows from sit-coms to talk shows. Lady Gaga is the queen of product placement in music videos. Check out her *Telephone* video - it has 56 million views at this writing. There are at least 11 products - see if you can pick them out.

Pros

- Creates top-of-mind awareness among customers/clients.
- People LOVE freebees - they'll tell their friends.
- Promotes goodwill.
- Can serve as a reward to loyal customers.
- Easy to link the item to the project, product or company purpose, ie. calculators from accountants, bottle openers from bottle beer company, etc.

Cons

- The promo item's quality can reflect on the branding.
- If not interesting or unique enough, most get tossed or put in a drawer.

References & Examples

http://www.aapglobal.com/aapstraps.php#vid

http://www.logo-lites.com/

When to Use

- To schmooze your best clients.
- As a thank you.
- As giveaway at event to spread brand exposure.
- As a collector's set.

Promotional Items

Category: Offline

Frequency: Regularly

What is it

Giveaways, swag and khak, are all names for promotional items. These branded items (logo'd) can be anything from magnets, pens, totebags, pins, keychains, can holders, sunglasses, calculators and disposable cameras all the way to watches, t-shirts, binoculars and golf balls, even mini message fans. There are virtually tens of thousands of products that you can put your logo on and give away to customers, prospects and event attendees as appreciation gifts or brand reminders. The more important the client, the more expensive and clever the swag should be.

Promotional items can also be given away individually to make a collector's set. Pins and coins are a good example. Either give away as bonuses after purchasing a certain level of product/service, or as a simple incentive for dropping into the location.

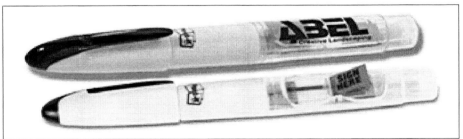

Photo: Captured at http://www.4imprint.ca/product/zoom.aspx?ca=C103364-T-H

Pros

- Can be picked up by all enabled cell phones.
- Bluetooth ad recipients can forward ad files to other cell phone users.
- Bluetooth hardware/software is inexpensive and no added costs.
- Gives street teams broader range with portable bluetooth servers.
- Low cost advertising.
- No cost to the recipients.
- No spam as recipients choose to be connected.
- Reaches phones in the immediate vacinity.
- Engaging for event and tradeshow attendees.
- Can custom program and change messaging quickly.
- Customer data collection and usability trends are big wins with RFID/NFC.
- Can send out customized and personalized messaging based on RFID/NFC user data.

Cons

- Privacy concerns may create a barrier to adoption of some of these technologies - particularly under the skin embedded chips.
- Possibility of technology being used unethically.
- Wireless Dynamics requires consumers purchase iPhone hardware until upgrade.
- Will take time for other smartphones to catch up.
- Bluetooth users can reject receiving ads.
- Not all people will or know how to activate their bluetooth.
- Still lots of consumer education needed to get this technology going.
- Still a way to go before the RFIC/NFC technology is unified enough to be adopted by consumers, merchants and advertisers.

Proximity Advertising

Category: Mobile, Offline

Frequency: Regularly, Adhoc

What is it

This is the future! At this writing the overall technology beyond contactless payments and inventory control is still being developed for advertising purposes. But it's getting started using mobile phones as the ubiquitous consumer tool to facilitate Bluetooth, RFID (Radio Frequency Identification), NFC (Near Field Communication) and WIFI.

It's incredible what you can do now. Bluetooth seems to be the furthest along with several companies selling bluetooth servers that can connect to any PC. With very little effort you can be sending out any number of graphic or data files to bluetooth enabled cell phones that pass within 100-200 yards/meters. Portable servers can even be purchased with battery life from 2 - 8 hours which are great for on-the-move street teams to deliver messaging to passers by anywhere. The downside is that cement walls tend to block the signal and the phones have to be manually set to receive bluetooth which will require consumer education. This technology is one-way in that an advertiser can send out info, but does not capture the data of the recipient.

RFID and NFC in its full form is a two-way technology that will give advertisers the ability to capture customer data while providing information/content to consumer NFC enabled mobile phones. As these phones pass establishments, billboards,

References & Examples

http://www.proximity-ads.com/PBMpresentation.html

http://www.proximity-ads.com/bluetooth-marketing.html

http://tinyurl.com/148Ways-RFID-AdWeek

http://www.wdi.ca/news.shtml#news12

www.mobioid.com

http://www.pressreleasepoint.com/potential-and-dangers-advertising-rfid

http://www.nearfield.org/2009/11/iphone-rfid-and-nfc-peripherals

http://deandonaldson.com/2009/11/06/rfid-equipped-iphone/

When to Use

- To capture user data.
- To target a very specific demographic.
- As a customer convenience.
- If you have a large base of cell phone using customers/prospects.
- To engage attendees at events and tradeshows.
- To draw in walk-by traffic.
- To optimize and enhance street team acvities.

Proximity Advertising Con't

Category: Mobile
Offline

Frequency: Regularly
Adhoc

Continued from page 247

and/or directly swipe a chip embedded product or poster, the phones will automatically stream dynamic content.

Wireless Dynamics, has created the iCarte attachment for iPhone/iTouch users while waiting for updated versions to be delivered with chips already embedded.

Another marketing use of RFIDs is in bracelets and embedded under the skin. Currently being used for high-security access purposes and for VIPs to gain special access to establishments and events. Users receive unique discounts with a simple swipe of the band. It's already happening. Great means to track purchases and facilitate payments.

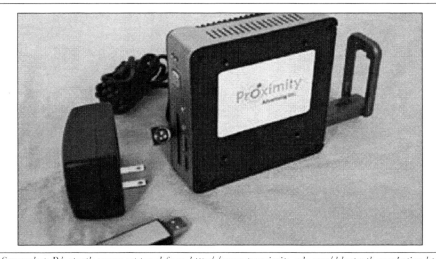

Screenshot: Bluetooth server captured from http://www.proximity-ads.com/bluetooth-marketing.html

Pros

- Publicity has more credibility than paid ads.
- It's free.
- Offline news outlets are always looking for interesting stories.
- Often easy to get coverage by local community news outlets.
- Online distribution services work in getting the message out and upping SEO rankings.
- Adding video or making the release an amateur video spot gives incentive for journalists to take the time to watch/read it.

Cons

- Can't control what is said/written.
- Takes people time to write releases, distribute and follow-up.
- Sending too many meaningless releases to offline media will be ignored.
- Free online services may require 2 weeks notice as they vet each release.

When to Use

- Electronic distribution through a provider is a must for your SEO strategy to increase rankings - use keywords in copy and include url.
- Only when you have something interesting (for others) to say.
- To announce partnerships.
- To alert media on special events or participation in a cause.
- Only if you have someone to follow-up after the distribution to offline media.

References & Examples

http://www.prnewswire.com/

http://www.pr.com/

http://tinyurl.com/148Ways-Flight-Release

http://tinyurl.com/148Ways-Mice-Release

http://www.stuntwomaninapencilskirt.com/

http://www.ShawneTV.com

Publicity

Category: Offline
Online

Frequency: Regularly

What is it

Publicity, also known as earned media, is when your company receives unpaid editorial coverage (written and/or image) of your business, product, service, event or activity. Publicity exposure is formated within the regular pages of newspapers, magazines; during the airtime of radio; anytime throughout television newscasts and talkshows; as well as mingled within the online world of search engines, blogs and related websites.

The most common way offline publicity is achieved is by emailing a media release (also called news or press release) or event alert to targeted media. The release must have a compelling angle or it will be ignored.

For online exposure, there are several service providers that will distribute your media release to search engines along with hundreds or thousands of targeted recipient sites. Online release distribution can be a major contributor to improved search engine rankings.

Photos: Taken by Charlene Brisson.

References & Examples

http://www.impressionsin-print.ca/cards/rackcards.php

When to Use

- To reach a specific demographic.

Rack Cards

Category: Offline

Frequency: Regularly

What is it

You see advertising racks in hotel lobbies, airports, tourist bureaus, shopping malls, community centres, libraries, restaurants, even at some local theatres. They hold dozens of standard sized promotional cards - usually 9" x 4", business card or post card size. The great thing about this is that most often the company that stocks the racks will also print the cards so you don't have to worry about keeping the racks stocked with your cards or storing inventory. Racks are usually targeted to specific segments like tourists, community and sporting events (runs), or arts and theatre.

Pros

- Inexpensive means to advertise.
- The distribution company keeps quantities of your cards on hand and will reprint as needed.
- Can use the printed cards for other distribution methods.

Cons

- Can be difficult to measure ROI.
- Has a high throw away factor.

Pros	Cons
• Can run a relatively robust radio campaign for the low cost of $2-$5k per week depending on the market and the dayparts you run.	• Expensive to run in highest listened to dayparts ie: morning (6 - 9 am) and afternoon (4 - 6:30 pm) drive.
• Can be low cost per spot outside of drive time.	• Fewer people listen to radio each year.
• The radio station will produce your spot for low or no cost and distribute to other stations you are advertising on.	• Young people from 14-30 years of age are almost exclusively plugged into iPods. Some iPods have FM radio - none have AM.
• Can target geographically and demographically.	• Unless something radical happens to change and promote this medium, AM radio won't exist within one generation - it is on its way out.
• Many radio stations will provide online bonus exposure at a reduced rate with campaigns.	
• A recent Nielsen study shows that 77% of people listen to a radio station at some time during their day.	• Some markets have no competition, it could be a one radio station market which makes it more difficult to negotiate pricing - yet again, these stations are usually pretty desperate for advertising, so the deals may be good to start with.

Radio - Commercial

Category: Offline / Online

Frequency: Regularly / Frequently

What is it

There are two basic commercial radio bands to advertise on - AM and FM. The choice of which band to advertise on depends entirely on who your target market demographics and geographics are.

Commercial radio uses advertising sales to make their profits and pay their bills. Radio commercials are called spots. Typically a spot runs in :30 or :60 (seconds). Some offer spots as short as :10 and :15. Radio stations differ in formats that appeal to different types of listeners. Formats can include rock, easy listening, all talk, soft rock, heavy metal, alternative, jazz, blues, etc.

Stations are divided into time slots and airtime is charged accordingly. The most listened to time slots are morning drive and afternoon drive. Stations are monitored by agencies like the BBM (Broadcast Bureau of Measurement) two to three times per year. They count the number of listeners per time slot through ballots that are sent out randomly to the general public.

Sponsorships of traffic reports, sports, the news and specific shows are available. There often involves a name intro, a :10 extro and a :30 ad.

References & Examples

http://tinyurl.com/148Ways-Nielsen-Radio

http://tinyurl.com/148Ways-Radio

When to Use

- For frequency - to reach your specific target market over and over again - the more they hear it, the more they will remember.

- As one element of a larger integrated campaign - an integrated campaign means running ads/messages in several different medium combinations such as radio, newspaper, online, direct mail, etc.

- To get the station to take on a promotional campaign. Most are free, however more often stations now insist on a major ad buy before they will give your promotional idea a go.

Radio - Commercial Con't

Category: Offline
Online

Frequency: Regularly
Frequently

Continued from page 255

Many stations, particularly talk stations run half-hour to one-hour programs, mostly on weekends, that are topic specific and fully sponsored. This means that the advertiser buys the entire show and provides all of the content. The advertiser either sponsors the show themselves by paying for the whole thing or solicits other advertisers to pay for the ads. This is quite common with financial, health and tech programs. It's similar to an infomercial but much less obvious as the radio station will often provide one of their announcers (paid by the sponsor) to host the program and will more than likely make it a live call-in show. Sponsor gifts for the first XX callers are almost always given away. Most effectively, the giveaways are promoted early and then actually given away later in the show to keep people listening.

Regular talk shows are terrific mediums to get interviewed on if you have a good educational story to tell about your product. See "Radio - Satellite" for more info about getting interviewed.

Screenshot: Captured at http://www.blogtalkradio.com/

References & Examples

http://www.targetspot.com

www.blogtalkradio.com

www,365live.com

http://www.spiritquestradio.com

When to Use

- To reach new audiences.
- When you have a message that can be turned into a good story.
- To educate people.
- To build relationship with the interviewer.

Radio - Online

Category: Online

Frequency: Adhoc

What is it

Online radio stations are available on almost every topic you can think of. Check out the Satellite Radio entry in this book to see how you can optimize your exposure on these stations. To find the stations that would be of value for you to be interviewed on - simply search "your topic/keywords" + "radio".

You can also target your advertising specifically to online stations or online streaming. Companies like http://www.targetspot.com have popped up to book advertisers (large and small) on internet radio stations. Worth checking into.

When researching who you want to spend your time and money with, make sure they have listeners and can prove it in some way. Don't believe everything you read provided by the producers. The best way is to measure response to your call to action (a giveaway of some kind).

Pros

- There are a lot of these stations popping up.
- No cost.
- Can easily test listenership by offering something free for the first XX people to call in.

Cons

- Never really sure how many listeners there are.
- Can be unprofessional.

Pros

- Free to create - only requirement is time, landline or skype connection, quality headset, computer, internet connection and knowledge of your topic.
- Have total control of the content.
- Great opportunity to educate yourself and team members through other industry leaders and suppliers through interviews.
- Can gain exposure to your guests' customers.
- Every time you run another show it provides one more reason to contact your customer and prospect database with something of value.
- Excellent way to create rich content for your website as downloadable podcasts; to add to a membership site; include in saleable information packages.
- Can use Twitter in real time to get people to send questions during the show.

Cons

- Can be very time consuming.
- Requires a knowledgeable and personable host.
- Consistency is critical - once you start, you can't stop.
- Some services don't offer live listener Q & A.
- Webinars are becoming more popular - where you can show visuals, and seem to be taking over a lot of audio.
- Have to use their URL for the free services, ie ... www.blogtalk.com/yourtitle.
- Services like www.365live.com claim to offer free radio to listeners - although I have yet to find one station that's free. When trying to listen live, they push their VIP listener membership starting at $5.95 per month. You decide.

Radio - Online Produce Your Own

Category: Online

Frequency: Frequently

What is it

You can create your own radio show using free online service providers like www.blogtalkradio.com or ones that charge monthly such as www.365live.com. Fees vary from as little as $60 per month to hundreds per month with limits on how many listeners you can have - so select wisely.

The great thing about producing your own show is that it offers you the channel to create rich content that you can put on your website as downloadable podcasts; use on your membership sites; and also create information packages that you may want to sell.

Use as a means to create loyalty and educate your customers and prospects on your industry and/or product. For example, if you're selling a product that helps Alzheimers Disease, your product should not be the focus of your content. Rather, your show topics should be about the disease itself such as ways to recognize Alzheimers; how to support the Alzheimers patient; interviews of experts that have new insights into Alzheimers, etc. Bring your product into the discussions when appropriate and perhaps with a :15 or :30 (second) spot each 15 minutes of the show (at most).

References & Examples

www.blogtalkradio.com

www.365live.com

http://www.spiritquestradio.com/hhp/hostinginfo.pdf

When to Use

- When you don't have a budget.
- To build credibility and establish yourself/your business as a leader in your industry.
- To educate customers and/or prospects by providing current info about and around your product/service.
- To create content for your website, membership site or info packages.

Radio - Online Produce Your Own Con't

Category: Online

Frequency: Frequently

Continued from page 261

Producing your own radio show will involve a pretty big learning curve and major commitment. It will require a great deal of time to book guests, prepare questions, pre-record, edit, post on your website and possibly convert into content that can be sold. It takes a team.

Screenshot: Captured from http://www.live365.com/index.live

Pros

- Radio interviews make terrific content for your website as free downloadable podcasts.
- Can include your interview into info programs you may be selling and post on your membership site.
- Several automakers are including satellite radio as original equipment which will increase the number of listeners/subscribers. Rolls Royce and Bentley include a lifetime subscription to Sirius Radio.

Cons

- Satellite radio didn't take off as big as planned - how many of your customers are listeners really?
- Nielsen reports Satellite listeners comprise only a small percentage of people listening to broadcast audio.
- It takes time to build relationship with the producers to get on their shows.
- Can be more difficult to pinpoint geographic targeting.

References & Examples

http://blog.nielsen.com/nielsen-wire/media_entertainment/within-ad-supported-media-broadcast-radio-reach-is-second-only-to-live-television-study-finds/

http://www.sirius.com/

http://www.siriuscanada.ca/en/

When to Use

- To build industry credibility and establish leadership in your industry.
- To create content for your website; membership site; and/or sales materials.

Radio - Satellite

Category: Offline / Online
Frequency: Adhoc

What is it

Satellite radio is 100% commercial free and supports itself through monthly subscription sales. Can also be known as pay radio or subscription radio. The most well known North American service is Sirius Radio brought to fame by shock jock Howard Stern and Oprah Winfrey. Sirius has 130 commercial free channels that appeal to pretty much any taste - news, entertainment, sports, music, talk. Users pay $6.99 - $19.99 per month and require a special portable or in-car receiver that can be purchased at any electronics stores like Best Buy or Future Shop for under $100. Sirius can also be subscribed and listened to over the internet. Muzak (background music in shops and on telephone hold) is subscription radio.

Satellite radio offers excellent opportunities to be interviewed on air about your topic. Talk shows are always looking for guests and news departments are looking for experts on topics that they can contact quickly. Audit the Sirius network for talk shows that relate to your product. Come up with an interesting angle, then pitch it to the producer. People pay much more attention to editorial than they do to ads anyway so if there's an audience, they'll get to know you and your product. Once you succeed with the first interview, try to secure a weekly spot - repetition is what advertisers rely on to seal the deal.

Screenshot: Captured from http://www.westhost.com/referral.html

References & Examples

http://www.westhost.com/referral.html

When to Use

- To generate new customers through word-of-mouth.
- To build up your database.

Referral Program

Category: Offline
Online

Frequency: Regularly

What is it

Similar to affiliate marketing, referral marketing is done on a more casual basis without a large investment in tracking software and training. Setting up a referral program can be as simple as giving a discount to the person who has made the referral after the new customer has purchased or taken advantage of your services.

Referrals can be generated by requesting outright; running a contest (provide XX referrals and win XX); or ad hoc when a new customer offers up a name at the time of sale.

Pros

- Easier to close a sale with people who have been sold to from their friends/colleagues.
- Customers feel good when they are rewarded for delivering a sale/prospect.
- Brings in new prospect data to sell to.
- Relatively easy to facilitate.

Cons

- Some businesses are uncomfortable to ask for referrals.
- Inevitably, some people will give invalid referrals for contests.

Photo: Jane Webb, owner of http://www.thebragcompany.com with Urban Rush hosts holding Brags, taken by Charlene Brisson.

Pros

- A high profile person's testimonial may significantly boost sales.
- Buyers trust peers before your company.
- Third party reviews are generally easy to come by online.

Cons

- You have to trust that the person giving the review and/or testimonial is in alignment with your company values. Two words - Tiger Woods.

References & Examples

http://www.cnbc.com/id/29961298

http://images.businessweek.com/ss/06/08/oprah/slideshow.htm

http://www.squidoo.com/how-to-pitch-the-oprah-show

http://reviews.cnet.com/

When to Use

- To generate confidence in your product/service.
- To place on sales letters, direct mail, email blasts, ecommerce pages, website and all collateral materials.

Reviews / Testimonials

Category: Offline / Online
Frequency: Regularly

What is it

Buyers will trust other buyers and third parties before they will trust your company. Presenting reviews and testimonials on your website and on collateral material will build customer confidence and provide a word-of-mouth opportunity for your product/service.

The reviews can be as simple as one line in quotes with the name and location of the source, all the way up to a full paragraph including photo of the source. It just depends on your space available, the significance of the reviewer and your goal. Buyers want to know that your company is legitimate; that you will provide the benefits you claim; and that others have had success with your product/service.

Securing a high profile person to provide a testimonial can result in a major boost in sales. An extreme example is the Oprah Effect. Companies have quadrupled their sales overnight through a simple mention on Oprah's show or a listing in her O Magazine. Nice to get Oprah's testimonial, but there are also many other high profile credible personalities that can provide you with testimonials. Also, don't underestimate the value of testimonials from regular people - they can be very powerful and the determining factor in making the sale.

Screenshot: Captured from https://www.starbucks.com/card/rewards

Pros

- Rewards programs increase share of wallet (amount person spends with your company).
- Builds customer retention.
- Can build a robust database.
- Simple programs can be manually executed and monitored if using pre-printed cards and stamps.

Cons

- At some point the rewards have to be paid out and many may come due at the same time.
- More complex reward programs require dedicated software to track points and redemption.

References & Examples

https://www.starbucks.com/card/rewards

http://tinyurl.com/148Ways-Barcode-Ikea

When to Use

- To encourage return customers - retention.
- To build a strong customer database.

Reward Program - Build Your Own

Category: Offline / Online
Frequency: Regularly

What is it

Generally, reward programs promote giving customers value points for purchasing. As points are collected, usually over several purchases, they can be exchanged for product and/or credit toward future purchases. Other means of rewards can include giving away one for every so many purchased. For example, coffee shops distribute coffee cards where a customer gets 1 free coffee after 5 are purchased.

Some clothing and product stores give reward cards that are stamped for every $50 purchase - when the customer has $250 in purchases, the customer gets 10% or $25 off their next purchase.

Reward programs give you the opportunity to collect customer data to build your database and to plot user trends. The rewards given away is the price you pay for the data. Of course, you can always solicit reward items from other companies in exchange for providing them with online and collateral exposure.

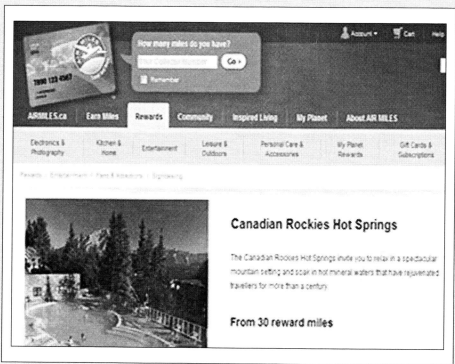

Screenshot: Captured at https://www.airmiles.ca/arrow/ProductTicketingDetails?productId=prod01586

References & Examples

http://tinyurl.com/148Ways-Rewards-Airmiles

http://www.loyaltyone.com/WhoWeAre/WhoWeAre.aspx

http://www.dbmarketing.com/articles/Art144.htm

When to Use

- To gain credibility with the program owner.
- To gain exposure outside of your customer base.

Reward Program - Provide Product

Category: Offline
Online

Frequency: Regularly
Adhoc

What is it

Other businesses' reward programs provide an opportunity for you to get your product in front of their customers. You can offer up your product to the program owner at a significantly reduced price in return for online and offline exposure in the program's overall marketing program. You may even be expected to provide your product/service for free in exchange for this exposure if the audience is really large.

Pros

- Cheaper to provide product than pay retail on advertising to get the same exposure.
- Usually included within a credible group of products/services.
- Introduces new people to your product/service.
- Only have to provide product when people redeem/exchange for points.

Cons

- Risk your product/service not being redeemed frequently.
- May be required to ship.
- May have to provide a large inventory.

Photo: Captured from http://www.rfidjournal.com/article/articleview/3005/

References & Examples

http://www.rfidjournal.com/article/view/3005

http://www.newmini.org/content/news.htm

http://www.geeksugar.com/Mini-Cooper-Launches-Talking-Billboards-I-Want-One-121948

When to Use

- To make a major impact within your market.
- To engage your customers.
- To increase retention.

RFID - Billboards

Category: Offline / Online

Frequency: Adhoc

What is it

This is wild! In 2007 Mini USA selected a group of 4,500 Mini Cooper owners in Miami, San Francisco, New York and Chicago for a pilot project using RFID enabled keyfobs (a piece that users put on their key ring). Recipients registered their keyfobs online providing some basic personal and job information. When these drivers would come within 500 feet of RFID activated Mini Cooper billboards, personalized messages, even birthday wishes, would flash in LED lights using the motorist's name and information.

Pros

- Excellent method to use with active clients.
- Creates massive PR about the campaign - media loves to talk about new media tactics.
- Creates an "I want one too" or "I'm not one of the cool people if I don't have one" following.

Cons

- Very expensive.
- Can take a lot of human resources to manage.
- Has to be set up in high traffic locations that those with the RFID embedded device travel.

Screenshot: Captured from http://smartchoice.tasterschoice.com/

References & Examples

http://www.startsampling.com/

http://smartchoice.taster-schoice.com/

When to Use

- To launch a new product.
- At events and tradeshows.
- To reach a new target market.

Sampling

Category: Offline

Frequency: Regularly

What is it

Sampling is a terrific way to get your product into your target markets' hands for a test without them having to make a financial commitment. Newspapers and magazines do this all the time by delivering their product for free to targeted zip/postal codes, or by giving a newspaper away with every coffee or breakfast sandwich sold. Food manufacturers do so at supermarkets. You can do sampling too in similar ways or by partnering up with another similar product with a like-minded audience. For example, if you have a maid service company, you can partner with a realtor and give a free one-hour cleaning to every new homeowner. It's great for the company you're partnering with as they get to provide added value to their customers. Be creative!

Pros
- Can piggyback off of another company's customer base.
- People love freebees so they're more open.
- Low risk to try.
- Can give a smaller quantity/value than what will be given at the full purchase price.

Cons
- If partnering, have to trust the company will represent your product/service respectably.
- Food sampling requires permits.
- Requires distribution channel.

Photo: Captured at http://www.cartartamerica.com/

References & Examples

http://tinyurl.com/148Ways-ShoppingCart

http://www.cartartamerica.com/

When to Use

- When selling a product/service available within the store.
- To increase suggestive sales.

Shopping Carts

Category: Offline

Frequency: Regularly

What is it

Shopping carts are a powerful medium to suggestive sell as customers roam the store aisles. The advertising message can be placed on the sides, in front, behind, inside on the bottom, on the child seat flap and along the handle.

Pros

- Provides brand visibility throughout the store to a very targeted audience.
- Encourages point-of-purchase sales.

Cons

- Not all grocery stores sell ad space on their carts. Your job is to convince them.
- Market exposure is limited to the people inside the supermarket.

Screenshot: Captured at http://www.qvcproductsearch.com/index.htm?rewrite=no

References & Examples

http://www.qvcproductsearch.com/index.htm?rewrite=no

http://tinyurl.com/148Ways-HSN

http://tinyurl.com/148Ways-ShoppingChannel

When to Use

- To move a lot of product.
- To get extreme exposure to the TV shopping market.
- To acquire national exposure.

Shopping Channel

Category: Offline

Frequency: Adhoc

What is it

QVC, Home Shopping Network and the Shopping Channel (Canadian) are terrific outlets to potentially sell massive amounts of product. After a series of interviews by product recruiters, if you're accepted, you'll be given a specific amount of airtime to test your ability to sell and the desireability of your product. The networks will provide a co-host to help you with your sales pitching and a coach. But when the camera goes live, you're on your own to present your product in the best light and to SELL. Each network has a baseline of how many pieces need to be sold within the test airtime for you to be asked back.

Pros

- Can move big quantities of product.
- Reaches a very targeted audience.
- Given opportunity to test sell a limited quanity of the product before making a large inventory investment.
- QVC and SC have product open house events.

Cons

- Takes a huge investment as you have to ship a large quantity of inventory to the network warehouse before it goes on sale.
- Limited product categories are accepted.
- Most only accept products, not services.

Photo: Captured at http://www.aarrowads.com/photo/come-virgin-megastore

References & Examples

http://www.aarrowads.com/about-us

http://www.wisegeek.com/what-is-sign-spinning.htm

http://www.signspinners-forapts.com

When to Use

- When opening a new location.
- Any place with heavy street traffic - cars particularly.
- When you have a very simple message to get across like a website.
- To point directionally to your location.

Sign Spinning

Category: Offline

Frequency: Frequently Adhoc

What is it

WOW - sign spinning is a real eye-catcher! Plus, it's simple. A person literally spins your sign around using breakdance type moves. Street corners, boulevards and anywhere that people congregate or pass by is the spin floor.

Could also be considered a human billboard, but is so creative and powerful, that it deserves a category all of its own.

Pros

- Sign spinning is unusual and can be a huge attention getter.
- A good spinner will draw a big crowd.
- Relatively inexpensive to produce a sign and hire a young person to do the spinning. Young, because it takes continuous effort and serious physical durability.

Cons

- Can be dangerous for those walking by and for the spinner who may accidentally end up in front of a car or aggressive scooter.
- May need a license in public areas.
- Spinning area must be large enough for the activity and possibly pedestrians.

Photo: Taken in Vancouver, BC by Andre Phaneuf.

References & Examples

http://www.digitalsignagemaker.com/en/home/index.asp

http://www.zonaeuropa.com/20051106_1.htm

When to Use

- To reach targeted audiences with high foot/car and transit traffic.

Signage - Dynamic LCD

Category: Offline

Frequency: Regularly

What is it

I love the streets of Hong Kong, New York City and Las Vegas - they have the most fabulous dynamic digital signage. Not just bigger than life dynamic signs in Times Square and on the strip, but also smaller signage in store windows and multi-screened displays in malls and other large venues. The signage is usually similar to a short commercial without the words and with minimal action. Elevators are also great places to run digital signage with the mini monitors up in the corner. The more technologically advanced we get, the more dynamic our signage becomes. Fabulous!

Pros

- Is still relatively new in most areas so will definitely draw attention.
- Has big impact if using multiple screens.
- The larger the screen, the greater the impact.
- Can be indoor or outdoor - not affected by weather.

Cons

- Can be very expensive to produce spot and rent in high traffic locations.
- If around too many other similar ads or screens, can lose impact.

Photo: Taken in Vancouver, BC by Charlene Brisson.

References & Examples

http://www.ledsignage.com/multi-line.htm

http://www.adaptivedisplays.com/

When to Use

- To draw attention to your establishment.
- To list your event taking place in venue.
- To reach a targeted high traffic area.

Signage - LED Text Only

Category: Offline

Frequency: Regularly

What is it

This LED signage is text only and scrolls from side-to-side or up and down, and often flashing. It looks similar to a giant lite-brite where the letters are made from dots of single or multi-colored lights.

These signs are run by computers and can be changed/updated rather quickly. They are often found on the front of convention centres, arenas, in corner store and restaurant windows and on busy street intersections. These signs really catch your attention at night.

Pros

- These signs are really visible in the evening darkness.
- Great at street corners when cars are stopped - lots of time to read the scrolling text.
- Can use multi-colored lights.

Cons

- Too much text will be too difficult to read.
- If the text moves too fast, also too difficult to read.
- Can be hard to read during the day.

Photo: Taken in Vancouver, BC by Charlene Brisson.

References & Examples

http://www.gstv.com/

http://tinyurl.com/148Ways-POPGrocery

http://www.ek3.com/digital-merchandising-products/ix-media.htm

http://www.ek3.com/digital-merchandising-products/digital-signage.htm

When to Use

- For point-of-sale advertising - when selling a product available at the location.
- To increase in-store, suggestive selling.

Signage - POP LCD

Category: Offline

Frequency: Regularly

What is it

Point-of-purchase monitors in most gas stations and super markets are becoming the norm. The monitor is often located at the cash-out and runs mini ads - looped at 10 or 15 seconds each promoting products that are available in the store - most specifically products that are available at check-out. www.GasTV.com locates monitors at the pump and intersperses ads with news, sports and weather programming. A pilot project is now underway to position monitors within the aisles of supermarkets.

New technology is intuitive and can be programmed to upsell based on realtime purchases. WOW!

Pros

- You have a mostly captive audience.
- Increases products sold near the cash out area.
- Can change messaging through website from offsite.
- Can be tied into coupon distribution and sales tracking.
- Can be programmed to upsell.

Cons

- Limited effectiveness as works best for products located within reach of cashout - like newspapers, chocolate bars, lotto tickets, muffins, coffee, etc.
- Can be quite expensive.

Photos: Taken in Vancouver, BC by Charlene Brisson.

References & Examples

http://www.displays2go.com/Category.aspx?ID=2257

http://www.aframesigns.us.com/

When to Use

- To draw walker, driver and transit rider attention outside store locations.
- When storefront and main signage is flat against building.
- Directional for events - particularly when repeating the event often.

Signage - Sandwich Boards

Category: Offline

Frequency: Regularly

What is it

From really sophisticated to hand-written chalk slates, sandwich boards are a terrific means to catch the attention of walkers, drivers and transit riders. Place them outside of your establishment, modern sandwich boards are light in weight (but heavy enough not to blow away) and can even feature brochure holders and changeable slots to easily and regularly change offers.

Tie a bunch of colorful balloons to your sandwich board and your location can't be missed.

Pros

- Very inexpensive.
- Can be very clever in the creative.
- Brochure/postcard holders work well - walkers will take away.
- Models you can write on are very flexible.

Cons

- May require approval of property owner to display.
- Some landloads won't allow sandwich boards on the property.

Photos: Captured at http://www.mediagraphics.net/smart-posters.php

References & Examples

http://www.mediagraphics.net/smart-posters.php

When to Use

- At events, tradeshows, and in show rooms.
- To draw people into a location.
- To present rich dynamic content.

Signage - Smart Poster

Category: Offline

Frequency: Regularly

What is it

Signage is everywhere. But static signage combined with an embedded screen streaming dynamic content (with or without sound) is very unusual. The screens can be integrated into signage as small as a piece of letterhead up to giant sized posters. Poster size can be flexible to fit screens that start at 3.5" and go to 32" in size.

This smart signage adds a twist to static signage that compels people to stop and watch.

Pros

- If content is interesting, can draw a crowd.
- Can update data very easily.
- Can loop existing ads and/or existing video content.
- Great sales tool to warm up prospects.

Cons

- Need to update fairly regularly.
- More expensive than static signage.
- Content and design has to look professional for it to work.

Photos: Taken in Vancouver, BC and New York City by Charlene Brisson.

References & Examples

http://www.giantpole.us/cat4b.asp?ID=25672

http://bannersonaroll.com

http://www.fastsigns.com/653

When to Use

- To attract attention.
- As directional.
- To create a "something's going on" energy.
- To identify location of product/service/booth.
- As welcomer at event.
- To announce product or special event.

Signage - Traditional

Category: Offline

Frequency: Regularly

What is it

Signs, signs, everywhere are signs. Hanging and stand-up signage comes in multiple shapes and sizes and is usually temporary. Creative is usually short messaging and bold graphics to attract attention and often, but not always, includes a URL. Commonly, traditional signage is used to bring attention to a product, service or company located in a specific area. Malls and super-markets will use signage at point-of-purchase to suggestive sell. Businesses hang banners off of the building to announce specials or grand openings. Also used at events and tradeshows for directional and information purposes.

Pros

- Very inexpensive.
- Brings attention to where your business/product/service is located.
- New fabrics and materials can improve durability and readability.
- Can create any shaped signage.

Cons

- Can be easily lost amongst other signage.
- Too busy creative and some materials/fabrics make reading difficult.
- Stand-up signage can be very fragile.

Pros

- Can get your message in front of a very targeted audience.
- Inexpensive to create.
- Easy to send out.
- Many list owners are negotiable on how to pay: by doing a list swap, or taking a percentage per sale (many want both).
- Often a list owner will provide a recommendation within your eblast.

Cons

- Not all lists are created equal - often you'll never know if the list you're getting is up-to-date or the number promised.
- If you don't set guidelines before making the deal, you'll have no control over what your list swap partner wants to send - insist upon final approval.
- Solo ad service providers' lists can be questionable.

References & Examples

http://www.soloadblaster.net/

http://tinyurl.com/148Ways-Solo

When to Use

- When you find a perfectly targeted ezine or company with the same target market that also has a large list.
- To test offers.
- To build your own opt-in list.

Solo Ads

Category: Online

Frequency: Regularly

What is it

A solo ad is a targeted sales email blast sent to a third party list. The blast is sent out directly to ezine or website opt-in subscription lists. Other than sending solo ads to your own list, you can pay service providers a flat fee to send out to thousands of their names (debatable how targeted they actually are). Or you can make deals with ezine list owners or companies that have similar products or services to yours. Offering to do a swap blast where you send their sales blast to your list will sometimes work as does paying a percentage of sales that result directly from the blast. It's worth paying 50% of the product sale price to the list owner depending on the price point as you get the info of the responder for your database.

Keep "list fatigue" in mind. Find out how often the list is being emailed, how old it is and what the average response is. Depending on your product and price point, you may be more successful by offering a free opt-in rather than selling a product directly through the blast.

Some companies and industries call solo ads joint venture (JV) blasts.

Pros

- Low to no risk for attendees.
- Most groups will let you sell your product at the end of the event.
- Groups are always looking for free speakers to fill meetings.
- By being seen as an industry leader, people will want to buy from you - if they like you.
- Can call on/identify people in the audience you know to build additional credibility.
- Can use to get leads by putting a draw box at the entrance.

When to Use

- To build relationship with like-minded targeted groups.
- To reach new targets.
- To build credibility within your industry.

Cons

- Have to be careful of how much you sell as it can turn people off.
- Non-targeted groups are a waste of time.
- Can take a lot of time to set up, travel, etc.
- Online technology can take a relatively sharp learning curve to manage.

References & Examples

http://www.bot.org/events/programs/business-workshop-series.aspx

http://www.keystonechiroinsac.com/custom_content/131723_lunch_and_learn_workshops.html

http://modern-baking.com/news/new_jersey_workshop/

http://www.gotomeeting.com/fec/webinar?Portal=www.gotowebinar.com

http:www.superconfernecepro.com

Speaking Engagements

Category: Offline / Online

Frequency: Frequently / Adhoc

What is it

Doing free presentations to targeted groups can be an excellent way to put a personality behind your product/service and reach new prospects. A content-rich, low sales presentation tied into a related topic can be very successful. For example, if you sell products to assist seniors - do a talk on preventing falls; if you sell financial services - do a talk on retirement savings, etc. Speaking opportunities are all around you - Boards of Trade, executive support groups, corporate lunch and learns, Chambers of Commerce, colleges and not-for-profit organizations. A 45 min to 1 hour presentation is all it takes. Be sure to put it up on your website that you/your company are available to do speaking engagements. If you have a shop location, offering free "how to" workshops can be real winners.

Online and telephone speaking engagements can also be a big draw for prospects. All it takes is a computer and/or a telephone and you can reach out globally. Relatively simple teleseminar and webinar service providers offer basic technology that you can easily and affordably access. Same principles apply to these engagements as face-to-face - rich information content and low sales.

Photo: Captured from http://www.summitmarin.com/raceteams.php

References & Examples

http://www.sports-city.org/sponsorship.php

http://www.summitmarin.com/raceteams.php

When to Use

- To associate with winners.
- To be seen as supporting the community.
- To gain massive exposure.

Sponsor a Team

Category: Offline

Frequency: Seasonal
Adhoc

What is it

Sports team sponsorship can cover anything from providing in-kind services or products to a team in return for association exposure; to paying for their uniforms to get title sponsorship (for small community teams); or paying large dollars to senior level teams that have several levels of sponsorship return including program ads, in-arena exposure like rink boards, on the score board, verbal acknowledgements, team member appearances, advertising rights and/or tickets to their big annual gala. Team sponsorships span from being relatively inexpensive for local teams to seven figures for professional sporting teams.

Pros

- Brand can receive significant media exposure.
- Can be considered as good corporate social responsibility.
- The more successful the team is, the more exposure you receive.
- Access to what many believe are the "heros" of society.

Cons

- The higher the profile of the team/sport, the more expensive the sponsorship.
- Risky, especially when athletes get involved in "unethical behavior" - as the sponsor you can be dragged into the controversy.

Pros
- Can usually negotiate more for your sponsorship level - ask for it.
- Builds a connection within an industry or program.
- Can give product/service instead of cash - called in-kind. I've also thrown in branded chocolates, cups, lanyards which saved the organizing committee money and gave my company terrific exposure. My favorite was using branded shoelaces for lanyards.
- A sponsorship almost always includes tickets - try to get VIP seating to network.

Cons
- The less the organization needs your money, the less they're willing to negotiate beyond the levels.
- Can become just one of many logos which I call logorama - need to decide if that's ok.
- Some in-kind sponsors get more exposure for giving less.

References & Examples

http://www.vancouver2010.com/more-2010-information/about-vanoc/sponsors-and-partners/vancouver-2010-sponsors/

http://www.prhelper.com/templates/sponsorship-package-1.php

When to Use
- When the sponsorship reaches your targeted audience.
- To show corporate good will within the community.
- To build credibility within an industry.

Sponsorships

Category: Offline / Online
Frequency: Adhoc

What is it

A sponsorship is when you agree to pay a pre-determined amount of money to event organizers in exchange for a specific amount of marketing exposure at and around the said event. Sponsorships are also available for almost anything that is public facing. Not-for-profits commonly solicit sponsorships for newsletters, programs and products.

Different levels of the sponsorships - often presented as Platinum, Gold, Silver, Bronze - have different exposure returns which includes everything from logo'd naming rights, onsite signage and website links through to distribution of sponsor materials and opportunities to connect directly with the audience through stage presentation at the event, sending email blasts and mailings.

Photo: Captured from http://www.businessinsider.com/america-is-over-let-the-new-century-begin-2010-2

Screenshot: Captured from http://www.squidoo.com

References & Examples

http://www.squidoo.com/squidoo

http://www.squidoo.com/lens-brainstorm

When to Use

- To increase your search engine rankings.
- To increase market and industry credibility.
- To drive traffic to your website.

Squidoo

Category: Online

Frequency: Regularly

What is it

A lens is specific to Squidoo.com. It is simply a platform to create a profile of yourself and/or what it is that you sell. In fact, Squidoo presents 49 different ways that you can create a lens so there is terrific flexibility in how you approach it.

The great thing about having a lens on Squidoo is that it will boost your search engine rankings (search engine optimization).

Pros

- It's completely free.
- Very easy to set-up.
- Adds to increased search engine rankings.

Cons

- Can become confusing if you also have a blog on your site - keep everything in a consistent form.
- Can lose a lot of time getting lost inside of Squidoo.

Photo: Taken in Vancouver, BC by Charlene Brisson.

References & Examples

http://newyork.backpage.com/BusinessServices/nyc-street-team-flyer-and-door-hanger-distribution-product-sampling-mobile-tours/9850901

When to Use

- To increase brand awareness.
- For product launches.
- To sample products.
- As part of a larger campaign.
- To target geographically.

Street Teams

Category: Offline

Frequency: Adhoc

What is it

A street team is a group of branded people hanging out in a high-traffic area distributing product samples, flyers and/or other promotional materials.

Predominantly working outdoors, street teams can also be seen inside trade shows, shopping malls, going from bar-to-bar and any other venue that lends itself to large attendance and promotional activities.

Pros

- Customizable per area or event.
- Gets product directly into hands of prospects.
- Generates energy around your product/service - looks like your brand is everywhere.
- Can get face-to-face feedback from customers and prospects.

Cons

- Can be weather dependent.
- Must provide very specific directions and supervision.
- Risk of team member presenting a poor image.
- Can be costly.
- Need permission to go into most establishments.
- Will be charged for access to some events and locations.

Photo: Captured at http://www.aapglobal.com/metrovista.php

References & Examples

http://www.aapglobal.com/metrovista.php

When to Use

- On very targeted transit routes.
- As part of an integrated campaign.
- For image advertising.

Subway Tunnel

Category: Offline

Frequency: Regularly
Adhoc

What is it

This is brilliant! A series of lightboxes are placed inside the subway tunnel walls and operate very similar to the first old fashion vaudville movies. Sensors pick up on the passing of the train and fills the train windows with an 8 second (or longer) brightly colored motion picture ad. It actually looks like the ad is dynamic. Entertaining for bored travelers.

Pros

- A very captive audience.
- Gives people something to look at while trying not to look at each other.
- Unique, so gets attention.
- Great repetition as people follow their daily rituals.
- Can include a text component so people can respond immediately.

Cons

- Can be difficult to measure ROI without direct response component.
- Expensive to execute.

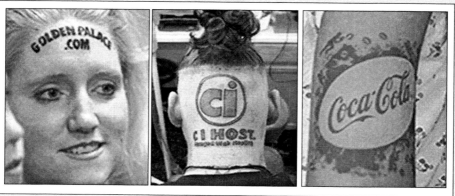

Photo: Captured at http://news.upickreviews.com/tattoo-advertising-for-money; Coca-Cola arm tattoo photo from Hong Kong Rugby 7's event taken by Charlene Brisson.

Pros

- People notice.
- Can get friends, family and/or employees to wear the temporary tattoos for your event - instant street teams!!!
- Handouts at events are really popular.
- Can be very inexpensive.

Cons

- A permanent tattoo means that whatever that person does will be linked to your product ... hmmm ... what behavior can you expect from a person that will wear a tattoo on the back of their head? Does it match your image?

References & Examples

http://www.tatad.com/sponsors.php

http://leaseyourbody.com

http://news.upickreviews.com/tattoo-advertising-for-money

http://tinyurl.com/148Ways-PermanentTattoos

When to Use

- To really make an impact.
- Great for optimizing event sponsorship.
- Excellent ambush marketing tactic.

Tattoos

Category: Offline

Frequency: Adhoc

What is it

You can have someone walking around with your logo (and sell line) permanently tattoo'd on their arm, chest, tummy, back or even on their forehead (yes forehead) or back of their bald head. Or if you're not in need of someone for a long duration, go for a group of people with temporary tattoos walking around your store, down the beach, at a large event - anywhere that your targeted customers are hanging out.

Another great strategy is to hand out temporary tattoos to event participants. Kids and lots of grown-ups love these. A few branded street team members walking around with a water spray bottle and a towel each can tattoo a lot of faces and arms over a couple of hours. It's kind of like face-painting - the line-ups will never end, and your brand will be everywhere.

Photo: Captured at http://www.miniatur-wunderland.com/exhibit/weekly-reports/year-7/article/weekly-report-no-290-cw-20/

References & Examples

http://tinyurl.com/148Ways-Team-EndCancer

http://www.vancouversun.com/2010sunrun/race_categories.html

http://tinyurl.com/148Ways-Team-BelowWaist

When to Use

- To reach a specific target audience that is also participating in the event.
- To show good corporate citizenship.
- To build leverage with the organizers.
- As a publicity angle.

Team Events

Category: Offline

Frequency: Adhoc

What is it

Each weekend there are dozens, if not hundreds of participation events organized around the country. Some, like many runs are simply for the accomplishment of it all. Others raise money and awareness for a specific topic. Almost all events accept team entries. This provides the opportunity to brand a team with your colors, logo and a fun name. Paying the entry fee and matching donations that your team members raise is great fodder for press coverge in smaller communities. If you don't have enough employees, invite friends and family to participate.

Pros

- Matching team donations can get your company media coverage.
- Builds terrific employee motivation.
- Builds relationship with the organizers.
- The more fun and outlandish your team, the more likely you'll get on-the-spot media exposure.
- There are so many events, your employees can find one they like.

Cons

- Really need a group of outgoing team members.
- Takes time - need one very organized person that will coordinate the team.
- May be forced to wear the race branded t-shirt.

Pros

- Easy for participants to join in.
- Long distance phone plans make it inexpensive for participants.
- There are free solution providers.
- Can record the session to replay later, post on your website or sell.
- Gives your prospect/customer extra value.
- Many service providers have weblinks for free listening.

Cons

- There is a learning curve to working the software.
- Live events are subject to technical difficulties - best to record.
- If over 500 participants, you may have to use more sophisticated software.
- Support from the software provider is rarely there when you really need it.

References & Examples

http://superconferencepro.com

http://freeaudioconferencing.com

http://www.nocostconference.com/

When to Use

- To provide education to your customers and prospects that will encourage them to buy your product.
- For list building.
- To develop content that can be sold later or put in sales kit.

Teleseminars

Category: Offline / Online

Frequency: Frequently / Adhoc

What is it

Similar to webinars, teleseminars are commonly one-hour content-rich information or educational sessions about a topic relevant to your product/service. Mostly used by information marketers, teleseminars can also be used by more traditional businesses as purchase bonuses or introductions to a higher priced, more thoughtful product purchase.

Usually teleseminars are one way communication by an expert or an interview of an expert, but some do include simple Q & A at the end. Most incorporate a soft-sell drive to a website/landing page for more information or to purchase a relatively low priced product or service. Great to offer a bonus at the end for those who stay through the entire call, but be sure to promote it at the beginning so listeners know about it. It's important that the call deliver content-rich information and not be all about selling your product. Selling is done at the end and in follow-up emails.

Teleconference solution providers, free and paid, make it relatively easy to deliver this marketing tactic. Some provide website domains to listen in on which removes the cost barrier for participants having to pay long distance fees.

Pros

- Can reach large audiences through specific shows.
- Most stations will negotiate on rates and throw in bonus spots if asked.
- Most TV stations can produce your spot and send it to other stations.
- Can post your spots on your website and YouTube.
- Can use your TV spot for pre-roll.

Cons

- Expensive to get into the game - an effective campaign starts around $20,000.
- TIVOs and PVRs can be set to skip ads.
- More and more people are watching TV on the internet without the ads.
- Frequency is critical, one spot is not enough.
- Can run too many ads and turn people off.

References & Examples

http://en-us.nielsen.com/rankings/insights/rankings/internet

http://ca.cision.com/assets/CA_Cision_Consumer_Rate_Card_2008.pdf

http://www.tvadvertisingsales.com/

When to Use

- As a component of a larger integrated campaign.
- For maximum impact during a high rated show.
- When you have a large budget.

Television

Category: Offline

Frequency: Regularly

What is it

Advertising sales is what commercial television stations use to pay their bills and make profit. In most hours, 12 minutes is available for sale in :15, :30 and :60 second spots (that's what they call commercials). Prime time is between 8:00 pm and 11:00 pm which is when the most number of people are watching at one time and therefore the cost of spots at this time are the most expensive.

In television, an advertiser pays for number of eyeballs watching. This determines a show's gross rating points (GRP) which are measured by Nielsen's Media Research. The more people watching, the more GRPs, hence the more expensive the spot. For example, the 2010 Superbowl. It delivered 106 million viewers - more than any show in the history of television to that date. Spots that ran during that Superbowl started at $2.7 million for :30 seconds. This fee does not include the production costs.

In the wee hours of the evening is commonly when you see advertisers with small budgets and often less than professionally produced ads. This is also the time when advertisers buy entire time blocks to run infomercials,

Screenshot: Captured from http://www.tvadvertisingsales.com/

Television Con't

Category: Offline

Frequency: Regularly

Continued from page 317

although we're seeing more infomercials running during the day than ever before.

Like radio, you are encourged to buy a package of spots to run during a variety of shows and time slots. Some of the ads will be categorized as ROS which means Run of Schedule in that they can appear at any time over a 24-hour period.

Reach and frequency are critical in successful campaigns. How much of the audience can you reach and how many times will they each see your spot? You can negotiate bonus spots with most stations. Some are also open to giving away products and/or services for viewer prizing during their own produced news and talk shows.

Also like radio, television offers sponsorships of its programming, traffic reports and weather reports which include visual and audio tags. Some stations also sell interstitials - up to :120 second spots which are basically long commercials that can be interpreted as programming - kind of like an electronic advertorial.

Photo: Taken in Vancouver, BC by Charlene Brisson.

References & Examples

http://www.inflatable2000.com/tents_pop.html

http://www.buyshade.com/

When to Use

- For branding exposure at events.
- To lend out to not-for-profits for their event use - and your free exposure.

Tent - Pop-up

Category: Offline

Frequency: Regularly
Adhoc

What is it

Branded pop-up tents are terrific for outdoor events and occasionally indoor trade shows. Using company colors and logo helps people find your location and brings together the branded elements of your business, product or service. With powerful creative, your pop-up tent can stand out among all others on site.

Pop-ups can be purchased with or without walls. The walls actually zip on and off for flexibility. In a pinch you can always rent a pop-up and hang a banner from the front and/or sides.

Pros

- Generally easy to set up.
- Can be used as a landmark in all promo materials - "look for the XX tent."
- Shields from the elements - rain, snow and sun.
- Solid creative can result in valuable branding opportunities.
- Many come with zip walls.

Cons

- No walls when it rains can be a challenge.
- Needs a decent size storage space.
- Large pop-ups take some muscle to set up.

Screenshot: Captured at http://www.freshplatform.com/casestudies/mobilezoo

Pros

- Using a good system, easy and quick to set up and low maintenance.
- Increased on-the-go deliverability when other methods aren't reaching your customer.
- Inexpensive means of communicating to your list.
- By opting in, customers pay for the text delivery fee.
- Can be used for cells and smartphones.

Cons

- Have to educate the customer that THEY pay for the message.
- Not all targets audience use cell phones.

When to Use

- Excellent method to integrate with email customer communications.
- For all transactional reminders.
- To announce offers, new products, flash sales or announcements.
- To engage customers with useful tips that are relevant to your product or service.

References & Examples

http://www.freshplatform.com/

Text Messaging

Category: Mobile

Frequency: Regularly

What is it

Text messaging is not just for teenagers. It's also gaining great strides in successfully communicating to customers. Text campaigns can be created for 1) triggered transactional texts (ie. your payment is due tomorrow/thank you for your order); 2) marketing efforts (ie. new offers, products and/or seasonal sales); and 3) message delivery (daily/weekly product tips or inspirational messaging). In all cases, customers have to opt-in to get text messages otherwise you'll be facing a lot of complaints and your campaigns will fail.

In some customer relationship management systems like www.salesforce.com, there are built in text messaging options that can easily be set-up. Other service providers offer text messaging platforms that you can sign up for on a monthly and per click thru basis.

Text messaging is also known as SMS (short message service).

Screenshot: Captured at http://www.ticketmaster.com/media/ticketwallets.html

References & Examples

http://www.ticketmaster.com/media/ticketwallets.html

http://www.zmginc.com/zmg/panels.html

When to Use

- To target very specific audiences.
- To show support of the arts or community event.
- For image advertising.

Ticket Jackets

Category: Offline

Frequency: Regularly

What is it

Also called ticket wallets, most boarding passes, concert and theatre tickets are presented to the customer in a folder (or ticket jacket) which more often than not, has advertising messages on the inside and back. In addition, the back of tickets themselves are commonly used as advertising opportunities.

Most ticket agencies, like Ticket Master, are hooked up with printers or have their own agency side to facilitate your advertising on their tickets and jackets.

Pros

- Less advertising competition since ad space is limited on the holder.
- Electronic ticket checkers let people keep their tickets in one piece.
- Can be seen as community supporter when on arts tickets (opera, theatre, ballet, etc.).

Cons

- In many cases, ticket holders are altogether being phased out.
- Electronic e-tickets are replacing mailings.

Photo: At F5Expo in Vancouver, BC taken by Charlene Brisson.

References & Examples

http://www.exhibitoronline.com/

http://www.trade-show-advisor.com/trade-show-promotions.html

When to Use

- To reach a targeted audience face-to-face.
- To secure real-time feedback from a targeted audience.
- For product launch.
- To generate leads.

Trade Shows

Category: Offline

Frequency: Adhoc

What is it
Great branding opportunity. A trade show is any gathering of companies in one place to promote products and/or services all within a specific topic or cause. Participation involves setting up a branded booth (standard is 10'x10') with professional signage. Most provide table and chair with rental. You pay extra for carpet, lighting, electricity, lights, plants and internet access. Professionally produced signage is critical for credibility.

Other ways of involvement in trade shows is to provide branded bags of any kind to the organizers to use for registration kit handouts or for participant handout collection. Other in-kind is often welcomed such as coffee cups/napkins (convention centres wouldn't allow this); branded volunteers; branded bottled water and/or door prizes.

Pros
- Big face-to-face reach in short period of time.
- Can get product directly into hands of buyers.
- Setting up lead generation activities at the booth can help build a targeted list.

Cons
- Number of attendees often don't meet the claims of trade show organizers.
- Can spend a lot of money for little return.
- Some exhibitors there only to do competitive research.

Photos: Captured from http://www.aapglobal.com/aapstraps.php#vid

References & Examples

http://www.aapglobal.com/aapstraps.php#vid

When to Use

- On very targeted transit routes.
- As part of an integrated campaign.
- As a component with other in-transit ads.
- When wrapping an entire transit vehicle - own it and brand it all!

Transportation Straps

Category: Offline

Frequency: Regularly Adhoc

What is it

All transit buses, subways, sky trains, metro and go trains have them - stablizers to keep stand up passengers from falling over during sudden starts and stops. Those with hanging straps also present advertising opportunities. Messaging can be positioned on the upper base of the strap. This would look particularly fabulous with a fully integrated brand - interior AND exterior. Why not take over the entire vehicle!

If vehicles don't have straps, and you have the budget, offer to purchase custom straps and have them installed.

Pros

- Ability to target geographically and demographically.
- Can make a strong impact with passengers.
- Can be part of a larger integrated theme within and without the vehicle.

Cons

- Not all transit authorities will accept this kind of advertising.
- Not all transit vehicles have the straps that can feature this advertising.

Pros

- Free to use.
- Great way for customers to keep updated.
- Easy to build a large base of followers.
- Great tool to provide links to further educate prospects.
- Can easily respond to customer service issues.
- Can schedule tweets to send at a later time.
- Lots of free software to support growing targeted followers and scheduling software.
- Have seen excellent results for networking.
- Easy to tweet from iPhone and smart phones.

References & Examples

www.twitter.com

http://techcrunch.com/2009/02/19/the-top-20-twitter-applications/

www.twellow.com

www.tweetdeck.com

http://blog.twitter.com/2010/04/hello-world.html

Cons

- Followers can be useless for sales if they're not your customer profile or someone you want to connect with.
- Many just try to get as many followers as possible - for no reason other than status.
- Direct selling on Twitter is frowned upon.
- If you can't be consistent, don't bother starting an account.
- Can spend a great deal of time getting lost in the Twitter world.

When to Use

- To increase online brand and SEO rankings.
- To provide engagement with customers and prospects.
- To announce last minute offers.
- To launch a new product.
- To drive people to your website.

Twitter

Category: Online, Mobile
Frequency: Regularly

What is it

Tweet! Tweet! Tweet! Twitter is one of the three most high profile and seemingly successful of the social media tools available to business owners (at the writing of this book). The basic idea is to acquire followers - ideally people who are interested in you and/or your product - and regularly post engaging 140 character messages.

Twitter is a relationship building tool that companies use for customer service, product announcements, providing educational links and general engagement. Because people tend to build large lists of followers for status, a good strategy is to drive customers and prospects to your Twitter account rather than relying list building within Twitter. Placing a Twitter link on your website, Facebook page, in emails and on printed materials will help to generate a targeted follower base.

At the time of this writing, Twitter is testing paid "promoted tweets" with multi-national partners like Best Buy, Bravo, Red Bull, Sony Pictures, Starbucks, and Virgin America, Universal Pictures, NBA, etc. On the right hand side of your Twitter home page check the bottom listing of the Trending Worldwide list. You'll see a trend listed with the word "promoted" immediately beside it highlighted in yellow. Once they work out the bugs I suspect you'll see it opened up to the rest of us.

Pros

- The more unique and loud the messaging, the more attention received.
- An entire fleet of branded cars makes huge impact.
- Even in down times, when not driving, the messaging is working for you.
- Very inexpensive when staffers pay for the lease which is usually less expensive because you get a bulk deal on leasing several vehicles.
- Tax write-off.
- Don't have to have staffers to make an impact, you can pay people to put your logo/messaging on their own cars.

Cons

- If this is your personal vehicle, then you are always on display.
- The vehicle/s must be kept spotless all of the time.
- Not all drivers will drive respectfully.
- In large spread out cities, it can be difficult to make an impact.

When to Use

- When you have a team of staff willing to drive a branded vehicle.
- As a delivery or customer service vehicle/s.
- When no one else in your area is doing it.
- In smaller communities for big impact.

References & Examples

http://www.wrapvehicles.com/

http://tinyurl.com/148Ways-Vehicles

Vehicles

Category: Offline

Frequency: Regularly

What is it

Every space on a vehicle can be used as advertising. From an entire wrap (see below) to bits and pieces like messaging on the doors, back and side windows, the trunk, the hood, the licence plate and frame. Even rotating hubcaps present an ideal promotional space. These are used widely in Singapore.

When not driving, try parketing. Coined by 1-800-Got-Junk, parketing is parking your branded vehicle in a high-traffic location when it's idle so that it is always working for you.

Photo: Taken in Vancouver, BC by Charlene Brisson.

Pros

- Gamer demographics are broader than you may think (http://www.theesa.com/facts/gameplayer.asp).
- Game developers are easy to find.
- Could sponsor an independent developer or a school's development project for a significantly reduced rate.
- Gamers spend at least one-hour a day on their games.
- Gaming competitions have become very popular providing additional sponsorship opportunities.

Cons

- Gamers can feel duped after entering their data to download or play the game and then get product spammed.
- Too much branding can be a turn off.
- May be difficult to measure true ROI.

References & Examples

hhttp://www.theesa.com/facts/index.asp

http://tinyurl.com/148Ways-VideoGames

http://www.iab.net/media/file/IAB-Games-PSR-Update_0913.pdf

http://www.advergame.com/

http://www.gamespot.com/news/6159158.html

http://www.candystand.com/

When to Use

- Primarily for large budget advertisers.
- As a free giveaway/promotional item.
- To generate leads.
- To gather user demographic and psychographic data.

Video Games - Advergaming

Category: Online, Mobile

Frequency: Regularly

What is it

According to the Entertainment Software Association, advergaming is the term originally coined by Wired Magazine to describe branded video games commissioned by companies to specifically promote their products. Before the advent of online webgames, floppy disks and CD branded video games were produced for many top brands to be used as promotional pieces.

Burger King entered the ring in 2006 by selling a series of three xBox games featuring "the King." Customers could purchase one for $3.99 with a BK value meal. Candystand.com, created for Nabisco, was the first free online game portal to appear.

According to the Casual Games Association, 400 new online games are launched each year and 200 million people worldwide are playing, spending $3 billion. The goal is to generate leads, market product and create word-of-mouth.

The most recent entry into gaming is the mobile phone, where brands are transforming advertising into mobile games.

Pros

- Your brand can be part of the game interactivity.
- Can reach a massive audience.
- Inside a game with gamers, you'll have a very focused audience.
- This audience spends most of their free time gaming so it may be one of the only ways to reach them.
- May be seen as being one of them which may translate into sales.
- Adds more realism to the games which Sony claims gamers want.
- Gamer demographs are broader than you may think. (http://www.theesa.com/facts/gameplayer.asp)

When to Use

- To go after a specific target group.
- If your product makes sense within the game.

Cons

- Gamers may be too focused to even notice the advertising.
- Online video games tend to come and go in terms of popularity.
- Can be difficult to measure true ROI.
- Some resistance from gamers to see branding in their domain.

References & Examples

http://tinyurl.com/148Ways-NascarGame

http://www.electronichouse.com/slideshow/category/4352/711

http://www.iab.net/media/file/IAB-Games-PSR-Update_0913.pdf

http://www.theesa.com/games-indailylife/advertising.asp

http://tinyurl.com/148Ways-emarketer

Video Games - In-Game

Category: Online, Mobile
Frequency: Regularly

What is it

Ok - admit it. How much time have you spent playing video games?

Advertising has finally "gone where no man (or woman) has gone before" as they say in Star Trek. According to emarketer.com, in-game advertising is expected to rise to $650 million annually by 2012. Advertising is embedded into the environment of a third party game in a number of ways such as 1) a McDonald's store on a street corner that the avatars travel; 2) rinkboards of a hockey video game or scoreboard on a baseball field; 3) billboards on the buildings 4) posters on walls 5) as tools used by avatars similar to product placement as is done on television and in the movies; 6) branded clothing to dress avatars in or 7) like in Everquest II where players need only type the word "pizza" in the browser and the Pizza Hut order screen appears - perfect for all night gamers that don't exit the game, even to eat. Players of Second Life have built entire interactive civilian constructs that include branded merchandise and stores strategically embedded by advertisers posing as players. Companies can purchase and create branded storefronts, create interactive experiences and sell merchandise using the site's custom money.

The most recent entry into gaming is the mobile phone where you can transform your advertising into a mobile game.

Screenshot: Captured at http://www.youtube.com/watch?v=smGmrpn2Vrk

References & Examples

http://www.youtube.com/watch?v=dMIO1cnzyzo

http://www.youtube.com/watch?v=smGmrpn2Vrk

http://www.prlog.org/tips/1048-free-video-press-release.html

When to Use

- When you have something of interest going on such as a business or product launch, promotional event, etc.

Video Media Releases

Category: Online

Frequency: Adhoc

What is it

Recording a live media release can be done simply with an inexpensive flip camera. Send the entire mp4 file to your media contacts on DVD or upload it onto YouTube and send the link. Be sure that the email you send it with has a compelling subject line. The email body should be short and to the point. So should the video - has to be less than 10 minutes to upload it to YouTube. Keep in mind that a media release really shouldn't be more than 3 minutes max.

Can turn the media release into a vodcast and upload it onto your website for viewing.

Pros

- The change of format from boring faxed-in releases provides a welcome relief to most journalists.
- Clips of your electronic release can also be used as-is on websites.

Cons

- If there's no transcription of the release, busy journalists won't transcribe for print - be sure to include a hard copy release.

Screenshot: Captured from http://www.altterrain.com/Projection_Media_Advertising_Hill_Holiday.htm

References & Examples

http://www.altterrain.com/light_projection_advertising.htm

http://www.projectionadvertising.co.uk/outdoor-projections.aspx

When to Use

- To launch a product.
- To make a statement.
- As part of an event.
- As an ambush tactic.

Video Projection

Category: Offline

Frequency: Adhoc

What is it

Ever been driving downtown in the evening and suddenly see a giant advertisement on the side of a building that's never been there before? That's guerilla video projectioning. Video projectionists usually set-up covertly in a backlane or parking lot so as not to draw attention and will change locations several times throughout the night.

Since it's obvious who's done the projecting by seeing the ad, permission from the building is usually acquired first, otherwise fines may be involved.

Pros

- Great guerilla tactic.
- Gets word-of-mouth going - can create some media talk in smaller centres.
- Usually can buy ad-libbed media talk on radio and/or TV to talk up the strange occurrences.

Cons

- Typically only short bursts.
- Difficult to measure ROI if no interactive component.
- Can be fined if not asked for permission from building owners/managers.

Pros

- Once it gets started, you don't have to do any work.
- Don't need 1 million viewers for a viral campaign to be successful - be clear on your goals.
- Assessment software is easy to use and very inexpensive.
- Videos can be created very inexpensively - the rougher the better.
- Social media is an effective platform for viral campaigns.

Cons

- Often difficult to tell what will go viral and what won't.
- No control - some viral campaigns started by customers can damage the brand.

References & Examples

http://www.cocacolazero.com/index.jsp

http://www.youtube.com/user/Blendtec

http://www.assessmentgenerator.com/

When to Use

- To generate lists.
- To drive website traffic.
- For product launch.
- To build positive brand energy.

Viral Marketing

Category: Online

Frequency: Frequently

What is it

Viral marketing is similar to word-of-mouth, but all done through the internet. If someone sees or finds something they like online, they send it to their friends, who send it to their friends, who send it to their friends, and so on and so on. Great viral marketing can make its way into millions of inboxes and social media pages around the world. All it takes is something interesting, fun or quirky. The trick however, is to capture the viewer's contact data.

Coke did a fabulous campaign called Facial Profiler on Facebook which finds your twin look-a-like on Facebook. Some companies run assessments with provocative topics like - "Do you have what it takes to be a millionaire?" or "Will you live to be 100?" Participants only receive answers back through email so they have to provide contact info. Inexpensive software like www.AssessmentGenerator.com provides an easy platform to create assessments.

YouTube is famous for facilitating enormous viral successes. A product example is Blendtec. They posted a series of 92 videos. The campaign, called "Will it Blend", shows their blender mashing up unusual objects from crowbars and iPhones to soft guns and video games. Some of their individual videos were viewed by over 7 million people. In total they have 95,635,651 views at the time of this writing.

Pros

- Relatively easy and inexpensive to create.
- Editing software is inexpensive.
- Lots of reasonably priced editors available.
- Gives credibility when customers can see the people behind the product/service.
- Can introduce audience to products through education.
- Keeps people coming back to your website.
- Using syndicates like iTunes provide credibility and distribution.
- Can insert sales messaging at front and back of vodcast like pre-roll video.
- The rougher the video, the better.

Cons

- Will take time to learn editing software.
- Just like podcasts, each recording must be interesting or people won't watch.
- Can be challenging to upload to iTunes, but worth it.

When to Use

- To show product/service "how to" and user reviews.
- To keep people coming back to your website.
- To build brand credibility.
- To educate customers and prospects.
- To target by interest.
- To soft-sell your products/services.
- As content delivery tool for membership sites, weekly tips, etc.
- As Q & A session recordings.

References & Examples

http://www.vodcasts.tv/vc.php?cid=12

http://www.freemarketingzone.com/rss/create-vodcasts.html

http://www.apple.com/itunes/podcasts/specs.html

Vodcasts

Category: Online

Frequency: Regularly

What is it

A vodcast, also called a video podcast, is a digital video file very similar to a podcast, but instead of audio, it's video. The video often contains information content for your customers and prospects. Posted on your website, usually in an mp4 file format, visitors can watch on your website or any YouTube-like service, or download off of your website onto their iPods. The maximum length of video to upload onto YouTube is 10 minutes. Syndication sites like iTunes, will accept your vodcast for distribution free of charge or to charge a fee. Be cognizant of size restrictions. You may have to break up a longer video into several short clips. Many broadcast networks provide their regular newscasts on vodcast for download.

Creating a vodcast has become relatively simple and inexpensive with the Flip Video Camera that plugs right into a USB port on your computer and downloads the video directly to YouTube. Alternatively, many computers now come with editing software. Adobe Premiere Elements is an easy to use editing software for PC's that you can buy for under $100. Mac's come packaged with an editing software.

Photo: Captured from http://www.onlinetel.ca

References & Examples

http://www.neptuneinnovations.com/vs.asp

http://www.centravoice.com/

When to Use

- To launch a new product or service.
- As an auto reminder of a payment due or to attend an event they signed up for.
- To announce a big sale or event.
- To generate new leads.

Voice Blasts

Category: Offline, Mobile

Frequency: Regularly, Adhoc

What is it

As long as you've captured phone numbers in your database, you can send customers and/or prospects a pre-recorded telephone message any time. It can be set-up so that the message is left only if the recipient answers the phone, only if they don't answer, or both.

With many service companies you can either do the recording yourself or have one of their voice professionals record the message for you.

Pros

- Increases payments and event show-ups.
- Easy to set up.
- Provides means for customer service outreach without human resources.
- Very easy to target specific customers and/or geographically.
- Can rent phone numbers from list brokers and/or telecommunication companies that offer the service.

Cons

- Lots of hang-ups on the first word without listening to the message.
- If the message is not authentic, it won't be listened to.
- Can be costly for large blasts.

Pros

- Very inexpensive when buying in bulk.
- Can be environmentally friendly with recycled bottles and having recycle bins near by.
- Not-for-profits love getting water donated for their outdoor/indoor events.
- People are very appreciative and see it as value since they usually have to pay more than $1.50 per bottle.

Cons

- Most hotels won't allow branded water on the premises.
- Can be seen as being environmentally unfriendly - particularly in light of the BP oil spill catastrophy.

References & Examples

http://www.ripplefxwater.com

http://www.logoh2o.com

When to Use

- As promotional giveaway at indoor and outdoor events.
- In lieu of sponsorship donation requests from not-for-profits.
- As handout to customers while waiting.

Water Bottles

Category: Offline

Frequency: Frequently Adhoc

What is it

Branded bottled water is one of the most effective ways to get your brand into the hands of customers at events, trade shows, conferences, and conventions. Also an excellent item to hand to customers while they're waiting.

Branded recycling bins (blue boxes) are necessary accessories to ensure that you're being environmentally sensitive - even in places that don't outwardly recycle - show, and make sure that you do.

Branding an existing product is called private labelling or white labelling.

Photos: Captured at http://www.ripplefxwater.com

Screenshot: www.dailymotion.com/video/x1dmfn_jeep-waterfall_ads

References & Examples

http://www.dailymotion.com/video/x1dmfn_jeep-waterfall_ads

http://www.youtube.com/watch?v=92dczT2OCno

When to Use

- To make a major impact.
- At tradeshows/events/conferences.
- For image advertising.

Waterfall Logo

Category: Offline

Frequency: Adhoc

What is it

Looking for something really unusual to wow the crowds at a trade show or conference? Turning a waterfall into your logo will certainly do the trick. The water is actually formed into the words and images through programmed valves. It looks absolutely incredible! You have to see it to believe it.

Pros

- A real attention getter.
- Will likely not be another display like it at the event.
- Can build it into a spectacular setting.

Cons

- Takes about 6 days to install and program.
- Difficult to measure ROI.
- Very expensive.

Pros

- Participants can sign-on for free to listen and watch.
- Relatively inexpensive to use software - as low as $50 per month.
- Most software allows you to do online registration, email communications and promotions.
- Can provide a phone number to listen for those unable to be at a computer.

Cons

- There is a learning curve to working the software.
- Live events are subject to technical difficulties.
- If over 500 participants you may have to use more sophisticated software.
- Support from the software provider is rarely there when you really need it.

References & Examples

http://www.gotomeeting.com/fec/webinar?Portal=www.gotowebinar.com

http:www.superconfernecepro.com

When to Use

- As a lead generator.
- To educate customers and prospects.
- To create content for your website or future selling.
- To post clips on YouTube which help in increasing your search engine rankings.

Webinars

Category: Online

Frequency: Frequently Adhoc

What is it

Online education is a powerful means of attracting prospects and building lists. With relatively easy-to-use technology like www.gotowebinar.com, conducting a one-hour, content-rich/low-sell online event can be executed with minimal technical ability. Webinars are a terrific means to share information on your product/service. You can do it yourself or interview experts within your industry to provide the information prospects are looking for.

Most webinar software is all inclusive and contains sophisticated (but-easy to-use) registration and email contact abilities to communicate with participants and promote to others.

It's often best to pre-record webinars, particularly if they involve any level of complexity, as it's so easy for things to go wrong. Spontaneity can be created by doing a 5 minute live interaction at the front end of the recording and perhaps a quick Q & A at the end. The interaction really depends on the expertise of the host, although there really shouldn't be any live interaction until you are confident in the technology.

What's terrific about webinars is that they can be replayed again and again as well as be sold as content at a later date.

Pros
- Runs 24/7 to provide information and sell.
- Inexpensive to put up a simple site.
- Gives industry credibility.
- Helps buyers and researchers find information on products and services.
- Gives customers an opportunity to send/receive feedback.
- Accessible world-wide.
- There are many tools and strategies to drive traffic to sites.

When to Use
- To sell online.
- To support offline business.
- To provide general or complex information for clients and prospects.
- For credibility.
- As a list generator.
- As a customer service tool.
- To educate.

Cons
- A poorly designed website affects brand and credibility.
- Servers sometimes go down making your site unaccessible.
- Information must be kept updated.

References & Examples

http://www.instantshift.com/2009/07/28/55-fresh-examples-of-corporate-website-designs/

http://www.2expertsdesign.com/web-designs/corporate-website-designs-by-wordpress

http://www.instantshift.com/2009/07/07/80-corporate-website-designs-for-design-inspiration/

http://www.widgetbox.com/mobile/make/

http://www.bmobilized.com

Website

Category: Online
Mobile

Frequency: Regularly

What is it

A website is the basic cost of doing business on the internet - like an online businesscard. It is the pages that appear when typing a URL/domain name at the top of your browser. A website is an information center that promotes your products or services to consumers who may not be within the store's reach or want to research their purchase before buying. Sites range from very simple brochureware that have general information and photos all the way to sophisticated interactive sales and marketing stores. Your website is the place to generate lists - give away information through a newsletter subscription or free whitepaper sign-up. If they've found your site, they're already qualified, so don't let them get away.

With iPhones and smartphones being used in record numbers to search for immediate information and browse websites, it's becoming critical for businesses to create a mobile-enabled site. Most web designers can do this when developing a new site. There are also service providers that you can use to easily reformat elements of your existing site within minutes. To see if your site is mobile enabled, simply do a browse on your smartphone.

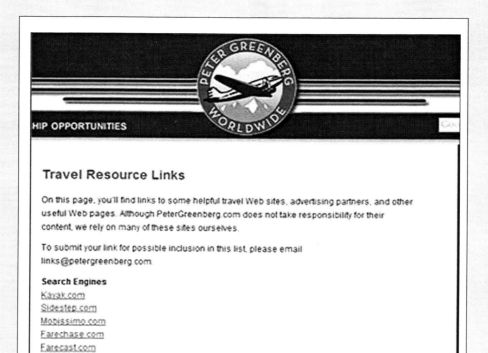

Screenshot: Captured at http://www.petergreenberg.com/?page_id=348

References & Examples

http://technorati.com/business/advertising/

When to Use

- To drive targeted visitors to your site.
- To gain credibility within your industry.
- To be seen as a valuable resource to your target.

Website Links

Category: Online

Frequency: Regularly

What is it

Called backlinks, getting your URL (link) on like-minded websites will provide a stream of targeted traffic to your site. In addition, links going into your site will help increase your rankings on most search engines, as part of search engine optimization (SEO). Like-minded sites are often looking for resources as well as information on products and services for their customers as it is seen as adding value to visitor experience. The link should also include a short write-up and logo if at all possible.

Pros

- Getting a link posted is absolultely free, although you may have to put theirs up on your site in return.
- Great for niche marketing.
- Helps improve search engine rankings.
- Can request more than a link, such as a listing, information paragraph and/or logo.

Cons

- It can take a lot of time to get links posted on other sites.
- Make sure the site your link is posted on is credible.

> **Sign Up Now**
> to Become a Member of GolfFitnessProducts.net and
> You'll Get FREE ACCESS to 5 Golf Fitness Reports (a $189 value!)
> that only Members can get. There's Nothing to Buy.
>
> All members also get 20% off Basic Golf Apparel!!
>
> **Membership Bonuses:**
>
> **FREE REPORT 1:** *"315 Seconds to Better Golf (Fitness Edition)"*
>
> **FREE REPORT 2:** *"7 Components Your Fitness Program Needs"*
>
> **FREE REPORT 3:** Dr. John Berardi's *"Gourmet Nutrition"*
> Sample PDF
>
> NEW! **FREE REPORT 4:** *"Treadmill for Golfers"* - A 6 week Hill Interval Program
>
> NEW! **FREE REPORT 5:** *"Eliminating Swing Faults"*
> from Mike Hansen
> Start Playing Better Golf NOW

Screenshot: Captured at http://www.golffitnessproducts.net/

References & Examples

http://whitepaperworld.com/

http://en.wikipedia.org/wiki/White_paper

www.getresponse.com

When to Use

- As a targeted lead generator.
- To position yourself/company as an industry or topic expert.
- To educate clients.
- To gain confidence from prospects and clients.
- To add rich content to your webiste.

Whitepapers / Reports

Category: Online

Frequency: Regularly

What is it

A whitepaper is an educational article written about a specific topic. The term whitepaper indicates that the author is an expert on the subject and that the article will solve a problem. Reports often include statistics and/or a collection of expert recommendations, steps and/or advice.

Offering a free downloadable whitepaper or report on your website is an excellent tool to use in exchange for name and email address (basic lead information). Only people interested in the topic will download it so you can build a very targeted list for future emailing.

Pros

- Attracts and builds a very targeted, topic specific list.
- Reasonably easy to set-up on website using email auto-responder software.
- Can access free whitepapers online to use on your site.
- Can provide site visitors a reason to navigate through your website.

Cons

- Takes time and effort to create valuable content.
- Need to keep the report up-to-date.
- Can take a fair amount of time to build relationship with list members before they buy - don't try to sell too soon - provide additional info.

Screenshot: http://commons.wikimedia.org/wiki/Category:Company_logos

References & Examples
http://commons.wikimedia.org/wiki/Main_Page

When to Use
- For when you want to provide high resolution photo/s of your product/service/personality/logo available for media, affiliates, etc. should they need it.

Wiki Media Commons

Category: Online

Frequency: Adhoc

What is it

People are always looking online for free photos to use on affiliate websites, design projects and social media posts. Why not have them use your branded photos. Upload high resolution images of your product/service/personality/logo on Wikimedia for free use for anyone to incorporate in their materials, websites, blogs, etc.

As of September 2009, 5 million pictures have been uploaded onto Wiki Media Commons.

Pros

- Helps SEO by placing your photos on the image page when searching Google and other search engines.
- You control the photos.
- Media and affiliates can easily access high resolution professional photos off of the internet that you want them to have.
- Can request to receive credit when used.

Cons

- Don't get your hopes up for voluminous downloads.
- Once posted, the photo is open game for anyone to use - then again, so is everything on your website so why not be pre-emptive.

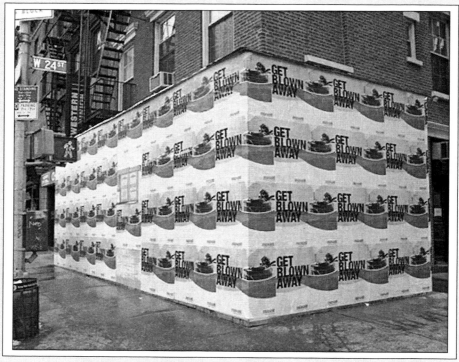

Photo: Captured from www.maxell-usa.com/getimage.aspx?id=23&type=ads

References & Examples

http://www.adsoutdoor.com/formatsSub.asp?id=8

http://www.streetteampromotion.com/wild-posting-national.html

When to Use

- To advertise events.
- To launch a new product.
- To target specific geographics.

Wildpostings

Category: Offline

Frequency: Adhoc

What is it

Wildpostings are large size posters that are placed side-by-side and one-on-top-of-the-other on the side of buildings, construction fences, barriers, walkways and windows that bring advertisers' messages to consumers at street level in a very public, in-your-face fashion - at least until they're stripped off by local municipalities or building owners. There are companies that will put up your posters in areas that have permission to do so and make sure they stay there.

Pros

- A series of the same posters in a row really attracts attention.
- Service providers will keep putting up posters torn down or graffiti'd.
- If creative is spectacular it can become a collectors item - creates a lot of talk.

Cons

- Can be torn down or vandalized.
- Most details only viewable by walkers-by, not drivers or bus riders.

Photo: Captured from http://www.altterrain.com/Wild_Postings_rip_away_posters.htm

References & Examples

http://www.altterrain.com/Wild_Postings_rip_away_posters.htm

http://www.matrixmediaservices.com/guerilla/wildpostings.php

When to Use

- To advertise events.
- To launch a new product.
- As a viral means to spreading your message.
- When you have a desirable image that everyone wants.
- To generate word-of-mouth.

Wildpostings - Rip Away

Category: Offline

Frequency: Adhoc

What is it

Similar to wildpostings except each posting is actually a pad of several posters so passersby can easily rip one off and take it with them. "Free Poster" printed at the top helps to let people know they are meant to be taken.

You will require a firm base and something to hook (zap strap) the pad to. Railings, chain link fences, any kind of scaffolding and even wooden telephone poles (where you can hammer in a hook or two) will work great.

Pros

- Can be hugely successful with the right image targeted to the right audience.
- Can turn into collector's items and on eBay.
- Great viral exposure as people repost in their offices and community centres.

Cons

- Have to replace them frequently if popular.
- Won't have impact if the image isn't cool enough or targeted to the wrong group.
- Many may end up in someone's apartment where they will never be seen.

Photo: Taken during the 2010 Vancouver Winter Olympics by Charlene Brisson.

References & Examples

http://denverwindowpainting.com

http://www.nolimitart.net/window-painting.html

When to Use

- Seasonal and special events.
- To participate in community activities.
- To be seen as supporting community events.
- As an attention getter.

Window Art

Category: Offline

Frequency: Adhoc

What is it

We often see window art during city events and at special times of the year. At Christmas, stores have their windows painted with Santas and reindeer. During the Winter Olympics, fabulous art displays showed up all throughout the City of Vancouver to celebrate, draw people and project energy. Mural window painters often go store-to-store offering to paint. Many painters are independents so if you have trouble finding a painter in your area try Craigslist or Kijiji.

Pros

- Catches attention of the walkers and drivers by.
- Adds to community spirit and local energy.
- Can incorporate your business and/or product profile within the art.
- Isn't permanent and can be washed off.
- Inexpensive.

Cons

- Not permanent.
- Can be difficult to measure the ROI.

Photo: Taken during the 2010 Vancouver Winter Olympics by Charlene Brisson.

References & Examples

http://www.millingtonassociates.com/

http://www.wowwindowsdisplay.com/

When to Use

- To engage customers, street and walk-by traffic.
- To participate in community activities.
- To be seen as supporting community events.
- To highlight product specials or seasonal excitement.

Window Display

Category: Offline

Frequency: Regular

What is it

Customer-facing windows provide the perfect opportunity to engage walkers-by. Using off-beat props intermingled with your products can create an artform that attracts attention and business. A window display is also an excellent platform to announce new products.

Pros

- Easy to do.
- Can recuit artists to provide props (for free).
- The more interesting the display, the more attention it will get.
- If tied in with community event, the display projects spirit.

Cons

- Has to be changed often to not become part of the woodwork.
- Can overload the display with props and people can't tell what you sell.

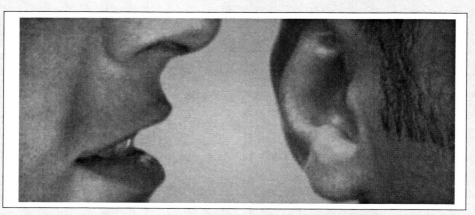

Photo: Captured from http://www.mediabistro.com/agencyspy/news/whats_your_word_worth_88260.asp

Pros

- Positive word-of-mouth starts with a good experience with your company which in most cases you can control.
- Word-of-mouth is considered more credible than paid advertising.
- Absolutely free.
- Can spread like wildfire.

Cons

- Negative word-of-mouth can have a major impact on sales.
- Very difficult to control word-of-mouth.
- Opinions are subjective and some people just can't be made happy.
- Can spread like wildfire.

References & Examples

http://www.joegirard.com/index.html

http://womma.org

When to Use

- To generate website traffic and increase sales.
- To build positive energy for your brand, product/service.

Word-of-Mouth

Category: Offline

Frequency: Regularly

What is it

The Law of 250 was coined by Joe Girard, author, and Guinness World Record holder as the World's Greatest Car Salesman since 1973. For Joe, sales is about building relationships. His law states that every person knows at least 250 people. So in dealing with a prospect or client, if they have a good experience, they'll tell 250 people, if they have a bad experience, they will tell 250 people (and that was before the invention of the internet).

A good product or service merits a recommendation from people who have used the product/service and are happy about it. In fact, the term word-of-mouth used to literally refer only to person-to-person exchanges of product/service experiences. But now, word-of-mouth advertising also includes sharing through blog sites, social networking status posts, Twitter, SMS, email, and other means of communication. Professionals who provide expert opinion on products/services also engage in word-of-mouth marketing.

Photo: By Charlene Brisson

Pros
- Puts your company up front with those who are not internet search savvy.
- Provides an alternative way of business search.
- YP is still a credible brand that many people recognize, trust and go to for listings.
- Can negotiate pricing based on number of cities and inclusion of spot color, etc.
- The internet does go down occassionally and the print editions are needed.

References & Examples

http://www.yellowpages.com

http://www.411.com/

Cons
- Can be costly for display ads in the printed editions that may not generate significant results.
- Printed Yellow Page books are being taken over by online search and directories.

When to Use
- For a target market that is NOT computer savvy.
- As a backup to internet listings.
- To maintain industry credibility.

Yellow Pages

Category: Offline
Online
Frequency: Regularly

What is it

It's amazing that every year like clockwork, a giant sized Yellow Pages book lands on my doorstep. Filled with thousands of business listings, this book is still used by millions of people in North American cities to search for products and services. The book has definitely reduced in size over time as the internet takes over from what was once the ONLY means to find a business by category.

Most companies that publish consumer business directories also offer an online directory to accommodate both the offline and online searching consumer.

There are two types of printed ads - listings and display. Listings are usually free - if the publisher is your phone company - with a charge for bolding and spot color. Display ads are typical in shapes and sizes to newspapers and usually offer only black and white and/or spot color options. To attract additional attention, prominent ads and creative advertising such as sticky notes are also available to be placed on the cover and spines of the book.

Online ad types are mostly listings with value plus charges for better placement, inclusion of photos and website addresses.

Screenshots: www.YouTube.com

Pros
- Relatively easy to set up a simple targeted campaign.
- Only pay for click thrus.
- Can start with a budget as low as $5 per day.
- Very targeted to appear ONLY to people who are searching for your keywords.

References & Examples
http://www.youtube.com/t/advertising_showcase

https://ads.youtube.com/pdf/YouTube-Promoted-Videos.pdf

http://www.youtube.com/t/advertising

http://www.youtube.com/t/advertising_self_service

http://www.youtube.com/threadbanger

Cons
- As of this writing, the YouTube Self-Service Ad portal is only available to US residents. Outside the US use a Google Ads Placement campaign to set-up keyword ads on YouTube.
- Can be a bit challenging to navigate the YouTube advertising space.

When to Use
- When you have interesting video content that provides a link through to where people can purchase your product or service.
- To build credibility and awareness within your industry.

YouTube - Ads

Category: Online
Mobile

Frequency: Regularly

What is it

With hundreds of millions of daily video views on YouTube, there's an enormous opportunity for targeted advertising. If you have a video worthy of promotion, you can get started immediately with a YouTube Self-Service Promoted Video Campaign which is tied in with Google Adwords and is based on keyword targeting.

Get your Google Adwords account set up first as the login is required to set up the YouTube ad program. Basic ad formats include text, display, video and in-video (the bar that shows up along the bottom :15 seconds into the video you're watching). More expensive formats include expandable click to play video ads, tandem ads, autoplay ads, masthead ads (banner ad on home page) and html in-video ads. Check out the showcase YouTube page.

YouTube also provides opportunities to pay to have your video placed on the home page for one day as a featured video; to run contests; become a content partner and/or co-launch or co-create programs. Paid keyword video searches are located at the bottom of the search list and are highlighted in pink.

Screenshots: www.YouTube.com/curebadbreath and www.YouTube.com/willitblend

Pros

- People love to see the people behind the product/service - show them.
- Including video buyer reviews on your website increases sales.
- Inexpensive to create videos - can use a Flip Camera and free editing software.
- Using the free YouTube brand channel provides one location where people can subscribe and watch ALL of your videos.
- Increases search engine rankings.

References & Examples

http://www.youtube.com/threadbanger

http://www.youtube.com/t/advertising

Cons

- Using humor can backfire.
- Others can post videos about your company that are unauthorized - check out the Dominos fiasco ...
  http://abcnews.go.com/Business/story?id=7355967&page=1

When to Use

- To build brand and product credibility.
- To educate people on issues around your product/service.
- To show a humorous side of your product/service.

YouTube - Videos

Category: Online, Mobile

Frequency: Regularly

What is it

Videos are an excellent way to interact with customers. According to Nielson Online in November 2009 Americans streamed (watched) 11,175,082,000 videos (http://tinyurl.com/yaf3wu6). Of those, 6.7 billion were streamed off of Youtube. As per daily views, as of January 2010 YouTube claims hundreds of millions of views per day. What an audience!

Just like video media releases, a YouTube video doesn't have to be high tech or a viral miracle, it can be simple and non-professional. Videos can and should be more than hard selling ads. In fact, if you use videos to provide rich information content and/or customer experiences with a product and/or service, the results can be excellent. Post videos on your own YouTube brand channel to drive people to one place on YouTube. It can be personalized as www.youtube.com/yourcompany. You can set up a channel within your YouTube account.

Place videos in emails, on websites, Facebook, Twitter and Linkedin to increase SEO rankings by increasing links to your website. Be certain to use keywords in your video discription and your URL.

Connect with Charlene on

 http://tinyurl.com/CharleneBrisson

 www.twitter.com/CharleneBrisson

 www.linkedin.com/in/CharleneBrisson

 www.YouTube.com/3StepMarketingPro

www.CharleneBrisson.com/Blog

www.3-StepMarketingPro.com

Chapter Seven
- Marketing Support -

16 Major Mistakes Marketers Make
and how to avoid them.

Find out what's keeping you from planning and executing campaigns that deliver optimum results.

Download this free ebook using **promo code: FREEBOOK8** at

3-StepMarketingPro.com

Fast-track your business to increased sales.

- **Plan Quickly, Execute Immediately**
  Over 2 days, you'll do all three proven steps and come out with an immediately executable marketing plan that **will** produce results.

- **A Repeatable System**
  You'll learn a simple system that you can use again and again for all future campaigns.

- **Access the Specialist**
  You'll have direct access to Charlene's 25 years of global marketing experience as she works with you to create a plan that delivers!

www.3StepMarketingBootcamp.com

Accelerate your Success.

- **One-on-One Strategic Results Program**
  Work with Charlene to achieve tangible sales and revenue results through joint strategizing using tried and true integrated marketing techniques and tactics. She can also work directly with your team.

- **Campaign Jump-Start Session**
  Don't require ongoing support? Get Charlene's one-time input on a new or existing campaign. She'll give you honest and experienced feedback including every strategy, tip, technique and secret that she knows to ensure your success.

3-StepMarketingPro.com

Each week a different specialist shares their marketing secrets.

- **Get the inside track.**
  Each week Charlene digs into one of the 148 tactics profiled in this book or a new tactics that she's discovered. You get the inside track on tips and results from experts in that week's featured marketing method.

- **Keep on top of new and emerging tactics.**
  You'll hear from the early adopters of new marketing tactics so that you can learn from their successes and decide what's right for your business.

- **Anytime Access.**
  Be on the live call to get your questions answered, but if you can't - each call is archived for later listening.

www.MarketingTacticTuesdays.com

Please share your tactics ...

There are hundreds of tactics and thousands of integrated combinations that you can use to get your message to market. I have done my best to include the most notable at the time of publishing, but the list is in no way complete.

I'd love to hear from you!

If you know of an advertising or promotional tactic not listed in **148 Ways***, or if you would like to add to any of the profiles, please email* **148Ways@charlenebrisson.com**. *I may include your suggestion in the next edition with a mention of your contribution. It may also appear on the www.MarketingTacticTuesdays.com members website.*

All the best in your marketing success,

Charlene Brisson
The 3-Step Marketing Pro

Book Charlene Brisson
for your next event.

Charlene delivers keynote presentations for beginner to advanced groups to show how effective marketing doesn't have to be complicated. Charlene is ...

- Dynamic • Engaging •
- Passionate • Entertaining •
- Motivational • Knowledgeable •
- Authentic • Skilled •
- and Fun •

Attendees leave inspired AND equipped to be better marketers.

Here's three of Charlene's keynote presentation topics:

Marketing ... Lady Gaga Style
Top 10 tips to market your way to the top of the charts!

It's Just Not that Complicated
Getting back-to-basics to blow away the competition.

From Marketing How? ... to Marketing WOW!
3 simple steps and 10 tactics to jump-start any campaign and get results.

To book Charlene for your next event visit

3-StepMarketingPro.com

Connect with Charlene on

 http://tinyurl.com/CharleneBrisson

 www.twitter.com/CharleneBrisson

 www.linkedin.com/in/CharleneBrisson

 www.YouTube.com/3StepMarketingPro

www.CharleneBrisson.com/Blog

www.3-StepMarketingPro.com